A. A. MOSCONA
DEPARTMENT OF BIOLOGY
UNIVERSITY OF CHICAGO
CUMMINGS LIFE SCIENCE CENTER
920 EAST 58TH STREET
CHICAGO, ILLINOIS 60637

Abnormal Embryogenesis:
Cellular and Molecular Aspects

Advances in the Study of Birth Defects
VOLUME 3

Abnormal Embryogenesis: Cellular and Molecular Aspects

EDITED BY

T. V. N. Persaud

University Park Press
Baltimore

Published in USA and Canada by
University Park Press
233 East Redwood Street
Baltimore, Maryland 21202

Published in UK by
MTP Press Limited
Falcon House
Lancaster, England

Copyright © 1979 MTP Press Limited

All rights reserved. No part of this publication
may be reproduced, stored in a retrieval
system, or transmitted in any form or by any
means, electronic, mechanical, photocopying,
recording or otherwise, without prior
permission from the publishers.

Library of Congress Cataloging in Publication Data

Main entry under title:
Abnormal embryogenesis.
   (Advances in the study of birth defects; v. 3)
   Includes bibliographical references and index.
   1. Abnormalities, Human.   2. Abnormalities (Animals)
3. Embryology, Experimental.   I. Persaud, T V N
II. Series.
QM691.A2   616'.043   79-16841
ISBN 0-8391-1470-2

Printed in Great Britain

Contents

List of Contributors — vii

Preface — ix

1 Ageing of spermatozoa and its effect on karyological abnormalities of resulting conceptuses
Patricia A. Martin-DeLeon and Evelyn L. Shaver — 1

2 The influence of teratogens on mouse embryo chromosome structure
R. G. Skalko, S. M. Tucci and R. H. Dropkin — 15

3 Interference with steps in collagen synthesis as a biochemical mechanism of teratogenesis
R. S. Bhatnagar and R. S. Rapaka — 29

4 Biochemical and ultrastructural aspects of [^{14}C]glucosamine utilization by normal and alkaloid treated preimplantation mouse blastocysts
Danica Dabich, R. A. Acey and Linda D. Hazlett — 51

5 Effects of cytochalasin B on preimplantation and early postimplantation mouse embryos *in vitro*
N. H. Granholm — 71

6 Pathogenesis of cyclophosphamide-induced fetal anomalies
D. Wendler — 95

7 The effect of retinoic acid on the developing hamster heart – an ultrastructural and morphological study
I. M. Taylor — 119

8 Physiological cell death in normal and abnormal rodent limb development
W. J. Scott, Jr. — 135

9 **Palate morphogenesis: role of contractile proteins and neurotransmitters**
E. F. Zimmerman 143

10 **Alterations in macromolecular synthesis related to abnormal palatal development**
R. M. Pratt, A. A. Figueroa, R. M. Greene, Ann L. Wilk and D. S. Salomon 161

11 ***In vitro* studies of virus-induced disorders of prenatal growth and development**
A. D. Heggie 177

12 **Impaired adaptation to extrauterine life: a teratogenic event**
R. De Meyer and M. Peeters 193

13 **Vital fluorochroming as a tool for embryonic cell death research**
B. Menkes, Oltea Prelipceanu and I. Căpâlnăşan 219

Index 243

List of Contributors

R. A. ACEY
Biochemistry Department
Wayne State University
School of Medicine
540 E. Canfield Avenue
Detroit, Michigan 48201, USA

R. S. BHATNAGAR
681 HSW, School of Dentistry
University of California
San Francisco, California 94143, USA

I. CĂPĂLNĂȘAN
Laboratory of Embryology
Center of Hygiene and Public Health
1900 Timisoara
Bv. Mihai Viteazul 24, Romania

DANICA DABICH
Biochemistry Department
Wayne State University
School of Medicine
540 E. Canfield Avenue
Detroit, Michigan 48201, USA

R. DE MEYER
Laboratoire de Tératologie et Génétique
Médicale, Départment de Pédiatrie
Université Catholique de Louvain
U.C.L. 5350 – Tour Pasteur
Avenue E. Mounier, 53
B-1200 Bruxelles, Belgium

R. H. DROPKIN
Department of Anatomy
Albany Medical College
47 New Scotland Avenue
Albany, New York 12208, USA

A. A. FIGUEROA
Craniofacial Development Section
Laboratory of Developmental Biology and
Anomalies
Bldg. 30, Room 405, NIDR, NIH
Bethesda, Maryland 20205, USA

N. H. GRANHOLM
Electron Microscope Laboratory
Department of Biology
South Dakota State University
Brookings, South Dakota 57007, USA

R. M. GREENE
Craniofacial Development Section
Laboratory of Developmental Biology and
Anomalies, Bldg. 30, Room 405
NIDR, NIH
Bethesda, Maryland 20205, USA

LINDA D. HAZLETT
Department of Anatomy
Wayne State University
School of Medicine
540 E. Canfield Avenue
Detroit, Michigan 48201, USA

A. D. HEGGIE
Department of Pediatrics
Case Western Reserve University
School of Medicine and
Rainbow Babies and Childrens Hospital
2101 Adelbert Road
Cleveland, Ohio 44106, USA

PATRICIA A. MARTIN-DELEON
School of Life and Health Sciences
University of Delaware
Newark, Delaware 19711, USA

B. MENKES
Laboratory of Embryology
Center of Hygiene and Public Health
1900 Timisoara
Bv. Mihai Viteazul 24, Romania

M. PEETERS
Laboratoire de Tératologie et Génétique
Médicale, Départment de Pédiatrie
Université Catholique de Louvain
U.C.L. 5350 – Tour Pasteur
Avenue E. Mounier, 53
B-1200, Bruxelles, Belgium

ABNORMAL EMBRYOGENESIS: CELLULAR AND MOLECULAR ASPECTS

R. M. PRATT
Craniofacial Development Section
Laboratory of Developmental Biology and Anomalies
Bldg. 30, Room 405
NIDR, NIH
Bethesda, Maryland 20205, USA

OLTEA PRELIPCEANU
Laboratory of Embryology
Center of Hygiene and Public Health
1900 Timisoara
Bv. Mihai Viteazul 24, Romania

R. S. RAPAKA
Food and Drug Administration,
HFD–S24 200 'C' Street SW,
Fed. Bldg., Room 6076
Washington DC 20204, USA

D. S. SALOMON
Craniofacial Development Section
Laboratory of Developmental Biology and Anomalies
Bldg. 30, Room 405
NIDR, NIH
Bethesda, Maryland 20205, USA

W. J. SCOTT, JR.
Children's Hospital Research Foundation
Elland and Bethesda Avenues
Cincinnati, Ohio 45229, USA

EVELYN L. SHAVER
Department of Anatomy
University of Western Ontario
London, Ontario, Canada

R. G. SKALKO
Department of Anatomy
East Tennessee State University
College of Medicine
Johnson City, Tennessee 37601, USA

I. M. TAYLOR
Department of Anatomy
Medical Sciences Building
University of Toronto
Toronto, Ontario, Canada M5S 1A8

S. M. TUCCI
Department of Anatomy
Albany Medical College
47 New Scotland Avenue
Albany, New York 12208, USA

D. WENDLER
Department of Anatomy
Karl-Marx-University
DDR 701 Leipzig
Liebigstrasse 13
German Democratic Republic

ANN L. WILK
Craniofacial Development Section
Laboratory of Developmental Biology and Anomalies
Bldg. 30, Room 405, NIDR, NIH
Bethesda, Maryland 20205, USA

E. F. ZIMMERMAN
Institute for Developmental Research
Children's Hospital Research Foundation
Elland and Bethesda Avenues
Cincinnati, Ohio 45229, USA

Preface

The study of birth defects has assumed an importance even greater now than in the past because mortality rates attributed to congenital anomalies have declined far less than those for other causes of death, such as infectious and nutritional diseases. It is estimated that as many as 50% of all pregnancies terminate as miscarriages. In the majority of cases this is the result of faulty development. Major congenital malformations are found in at least 2% of all liveborn infants, and 22% of all stillbirths and infant deaths are associated with severe congenital anomalies.

Teratological studies of an experimental nature are neither ethical nor justifiable in humans. Numerous investigations have been carried out in laboratory animals and other experimental models in order to improve our understanding of abnormal intrauterine development. In less than two decades the field of experimental teratology has advanced phenomenally. As a result of the wide range of information that is now accumulating, it has become possible to obtain an insight into the causes, mechanisms, and prevention of birth defects. However, considerable work will be needed before these problems can be resolved.

The contributions in this volume include some of the more recent and exciting observations on the cellular and molecular aspects of developmental defects. It is not only a documentation of the latest experimental work, but it also indicates new and important areas for future research.

I am most grateful to the distinguished panel of contributors. Their enthusiasm and cooperation have made this volume possible. My sincere thanks are due to the publishers, especially Mr D. G. T. Bloomer, Managing Director, MTP Press Limited, for their encouragement and for extending to me every kindness. Finally, I am much indebted to my secretary, Mrs Barbara Clune, who has lightened the burden of editing this book.

Winnipeg, Canada T. V. N. Persaud
September, 1979

1
Ageing of spermatozoa and its effect on karyological abnormalities of resulting conceptuses

PATRICIA A. MARTIN-DELEON AND EVELYN L SHAVER

INTRODUCTION

Spermatozoa undergo degenerative changes with the passage of time during storage prior to fertilization. The effects of these changes have been reviewed by Salisbury and Hart[1] and more recently by Vander Vliet and Hafez[2]. The ageing process or senescence, which seems to be a universal one, may occur in the male or female reproductive tracts as well as *in vitro* after prolonged storage at different temperatures. Although the effects of ageing vary according to the environment in which the ageing occurs, a loss of fertilizing capability seems to be common to all environments.

Embryonic development can, however, be initiated by defective as well as physiologically fit spermatozoa. With defective spermatozoa of several experimental and farm animals, a large proportion of the oocytes fertilized do not develop through the fetal stages[3]. Guerrero and Rojas[4], studying the probabilities of abortion after insemination on a given day of the menstrual cycle in relation to the day of the shift in the basal body temperature, concluded that an increased frequency of spontaneous abortions is associated with spermatozoa aged in the female tract. The study was conducted on 965 patients visiting family planning and sterility clinics on a regular basis. In addition to fetal loss, there is evidence in the rabbit that congenital defects may result from the fertilization of oocytes with sperm aged in the male tract[5].

In man[6] and other mammals the occurrence of heteroploid chromosome complements in embryos is perhaps the most important cause of abortion or embryonic death. Evidence that fertilization of aged oocytes results in a significant increase in the incidence of chromosomal abnormalities[7-9] has stimulated an interest in spermatozoan ageing and its effects on chromosome anomalies in resulting embryos. This chapter will review cytogenetic and cytological studies in vertebrate embryos resulting from spermatozoa aged in different environments.

DEVELOPMENTAL AND CHROMOSOMAL ABNORMALITIES FOLLOWING SPERM AGEING IN THE MALE TRACT

In the male genital tract, spermatozoa may be aged after prolonged periods of sexual rest or experimentally by ligation of the corpus epididymides, a procedure which prevents the entry of fresh spermatozoa into subsequent ejaculates.

Studies have shown that spermatozoa aged in the male tract lose their ability to contribute to normal embryogenesis before losing their fertilizing capability. Young[10], working with the guinea pig, demonstrated that spermatozoa retained in the epididymides for 20–25 days resulted in at least a fivefold increase in the incidence of embryonic and fetal death as compared to controls. In the rabbit, Igboeli and Foote[11] reported a decrease in cleavage and kindling rates when spermatozoa had aged 28–35 days in the male. After isolating spermatozoa in the caudal epididymides and vas deferens by ligating the corpus epididymides, Tesh and Glover[5] observed increased pre- and post-implantation loss with increasing spermatozoal age. Several fetuses removed from the uterus near term had abnormal skull sutures and lacked gall bladders.

In order to provide further information on the causes of the abnormal prenatal effects in the rabbit, Martin-DeLeon and co-workers[12] undertook chromosomal studies on six-day blastocysts resulting from the fertilization of oocytes with spermatozoa aged in the male tract. In this study, ageing was accomplished by techniques similar to those employed by Tesh and Glover[5]. Females were inseminated with spermatozoa aged in eight males for 7–35 days. Control data were obtained by insemination of females with ejaculates from each male 2–4 days prior to isolation of the caudal epididymides and sham-operated males. There were large numbers of inactive spermatozoa in ejaculates with spermatozoa aged at least 21–35 days. Ejaculates collected 14 days after isolation of the caudal epididymides contained a large number of decapitated spermatozoa and a number of head cap and tail defects. The main defects seen after 7 days were head cap abnormalities.

Spermatozoa retained in the caudal epididymides for 35 days resulted in no fertilized oocytes that developed to six-day blastocysts. Retention of spermatozoa for 7 to 27 days resulted in an 11% incidence of chromosome anomalies, whereas the incidence found in the 125 blastocysts in the control group was 0.8%. The difference in these frequencies was significant at the 1% level. The most frequent abnormality was trisomy with a modal chromosome complement of 45. Oocytes fertilized by spermatozoa aged 12 or more days gave rise to five blastocysts with anomalies, three of which were trisomic for autosomes. The other two abnormal blastocysts were a 44/45 chromosome mosaic and a blastocyst with a deletion of the short arm of a homologue of chromosome 5. Spermatozoa aged at least 7 days gave rise to three abnormal blastocysts; one triploid (66,XXY) and two 2n/4n mosaics each with an XXXX gonosomal complement. The sex ratio of all blastocysts in the study was unaffected by ageing sperm.

Later, in a similar study using the identical procedure to the one described above, Tesh and co-workers[13] recorded "a degree of polyploidy" in rabbit blastocysts resulting from aged spermatozoa, whereas no chromosomally abnormal blastocysts were found in the control series. This study therefore confirms the findings of Martin-DeLeon and co-workers[12].

The qualitative aspects of chromosome anomalies are important when considering the fate of the embryos in which they are found. Although any heteroploid condition will tend to impair normal development, some conditions have more severe consequences. In general, the types of abnormalities produced in aged sperm in the above studies[12,13] have been shown to exert extremely severe effects on embryogenesis. Triploidy in the rabbit[14], as in humans[6], is lethal to the developing embryo, whereas trisomies for autosomes and deletions are associated with abortions and gross malformations in man. It seems, therefore, that chromosome aberrations would account for at least a proportion of the embryonic mortality and morbidity seen when spermatozoa are aged in the male tract[5,10].

From the limited work produced so far it is clear that this area requires further investigations using different species and examining early and later stages of gestation for phenotypic and chromosomal anomalies. Recently, a developmental and karyological study of pre- and postimplantation mouse embryos resulting from sperm aged in the male tract has been initiated[15]. Modern chromosome banding techniques to enhance the detection of minute structural anomalies are being employed in this study. In addition, attempts to elucidate the mechanisms involved in the origin of any resulting triploid blastocysts will be made. For this, strains of mice with different C-band polymorphisms serving as markers will be used to identify the genomes of contributing gametes.

DEVELOPMENTAL AND CHROMOSOMAL ABNORMALITIES FOLLOWING SPERM AGEING IN THE TRACT

Spermatozoa may be aged in the female tract by prolonging the interval between insemination and ovulation. Early studies[16,17] indicated that the incidence of embryonic anomalies in fowl increased with advancing age of spermatozoa. In mammals, the effect of *in utero* sperm ageing on embryonic development has been thoroughly tested in the rabbit[18-20] where artificial insemination and the induction of ovulation by hormone injection are easily carried out.

That both pre- and postimplantation loss contribute to a decline in fertility when *in utero* aged spermatozoa are used to fertilize oocytes was first shown by Tesh[18]. The losses appeared to occur at progressively earlier stages with increasing spermatozoal age. Similar findings were obtained by O'Ferrall[19]. A reduction of cleavage rate[20], a marked increase in dispermy[21] and digyny accompanied by abnormal male pronuclear development[22] have been shown to occur in oocytes fertilized by rabbit sperm aged *in utero*. The latter phenomena, dispermy and digyny, both arise from anomalies of fertilization.

Since fertilization and cleavage errors are two major causes of chromosome anomalies arising in embryos, it was of interest to undertake cytogenetic studies of embryos resulting from the fertilization of oocytes by *in utero* aged spermatozoa. The chromosome complement of blastocysts recovered from rabbits in which ova were fertilized by aged sperm was studied by Martin and Shaver[23]. For this study females were artificially inseminated then induced to ovulate with an injection of human gonadotropin at various times ranging from 0–21 h postinsemination. This allows sperm to be retained in the female tract for 10–31 h before ovulation, assuming ovulation occurs 10 h postinjection. The recovery rate, which is the percentage of the blastocysts recovered compared with the number of corpora lutea in the ovaries, peaked when oocytes were fertilized by spermatozoa retained for 18 h in the female tract before ovulation. After this period there was a decline in the recovery rate leading to a complete failure to obtain blastocysts when sperm were retained for 31 h. The cytogenetic analysis of the blastocysts revealed a significant increase in the frequency of chromosome abnormalities in blastocysts resulting from fertilization of oocytes by aged sperm as compared to controls ($p < 0.01$). The anomalies were distributed throughout the various ageing periods and were of two major types, mosaicism and chimerism. The mosaics which were the most frequent (77%) could be subdivided in diploid/tetraploid mosaics (euploid series) and $2n - 1/2n$ or $2n + 1/2n$ (aneuploid series).

Martin and Shaver[23] postulated that the mosaic blastocysts could have arisen through errors of cleavage divisions in cells derived from a single zygote. It is possible that this impairment of mitosis perhaps results from abnormal male pronuclear development in aged sperm. Chimeras, unlike mosaics, have cells arising from two or more distinct zygote lineages. The three chimeric blastocysts contained about equal numbers of cells with an XY and an XX sex chromosome complement. The origin of the chimeras was thought to be through dispermy involving the ovum pronucleus and the second polar body[23].

The effects of spermatozoa aged in the female reproductive tract on chromosome aberrations in early developing chick embryos have been studied by Lodge and associates[24]. Ageing of sperm in the chicken as in the rabbit resulted in an increased frequency of chromosome abnormalities. The incidence of aberrations in fertile eggs collected on days 2–4 following insemination was 0.458% as compared to 2.2% in those collected on days 10–15. The authors concluded that chromosome aberrations would account for about 25% of the increase in embryonic mortality from *in vivo* sperm storage. The anomalies included 43% mosaics, 29% triploids and a haploid and a monosomy each accounted for 14%.

It is of interest that in the chicken as in the rabbit the most frequent anomaly seen was mosaicism. Mosaicism which occurs to a limited degree in human abortuses[25] is generally compatible with extrauterine life. Similarly, sex chromosome chimeras do not contribute to the load of chromosome abnormalities seen in spontaneous abortions. The developmental defects seen in chimeras are mainly confined to the reproductive system. Most human chimeras are mildly affected individuals and are not usually diagnosed before puberty.

As in the case of sperm ageing in the male tract, additional studies in different species should be done to determine the frequencies and types of chromosomal abnormalities induced by sperm ageing *in utero*.

DEVELOPMENTAL AND CHROMOSOMAL ABNORMALITIES FOLLOWING SPERM STORAGE *IN VITRO*

Semen from a variety of species has been stored at temperatures either above or below freezing. However, the reports are conflicting concerning the fertility of spermatozoa following *in vitro* storage. An ageing effect has been postulated by Salisbury and Hart to occur when spermatozoa are stored *in vitro*[1,26]. Fertility was found to decrease and embryonic mortality to increase in cattle following increased duration of sperm storage. Other investigators, however, have not observed a decline in fertility, particularly when spermatozoa were stored at $-196\,°C$[27,28]. It is difficult to compare the various studies because of the differing techniques, conditions and duration of storage and the criteria used to define fertility.

There have been relatively few studies in which the conceptus arising from stored spermatozoa was examined either during the pre- or postimplantation period. The following sections will deal with the development of the conceptus and studies that have examined it directly.

In vitro storage above 0 °C

The strongest evidence that an ageing effect occurs *in vitro* is from data derived following storage of spermatozoa at temperatures above freezing. A decrease in motility[29,30] and an increase in the number of dead spermatozoa[31] have been shown to occur during storage. Fertility, however, was found to increase during the first 2 or 3 days of spermatozoan storage at 4–5 °C then decline[31–33]. This fertility peak was attributed to the changing relative ability of stored sperm to effect fertilization[1]. Abnormal spermatozoa competed effectively with normal sperm for fertilization sites, initially. Later they were unable to do so and more oocytes were fertilized by normal spermatozoa, leading to the increase in fertility.

The decrease in fertility that occurred after the fertility peak could have been the result of the failure by the spermatozoon to penetrate the oocyte as well as death of the zygote. The fertilization rate obtained in several studies

Table 1.1 The effect on fertilization rate of sperm stored *in vitro* at temperatures above 0 °C

Species	Storage temp. (°C)	Duration of storage (h)	Fertilization rate (%)		Source
			Fresh sperm	Stored sperm	
Mouse	22–25	24	65.2	14.9	Bentiz and Graves, 1972[34]
Rabbit	21–27	144	95.0	8.4	Gulyas, 1968[29]
Rabbit	5	48	92.7	73.2	Koefoed-Johnsen *et al.*, 1971[35]
Rabbit	5	120	87.0	44.0	Maurer *et al.*, 1976[30]
Swine	6–8	54	61.0	39.3	First *et al.*, 1963[36]

using mouse[34], rabbit[29, 30, 35] and boar[36] semen showed that the percentage of cleaved ova, recovered from females inseminated with spermatozoa stored until the end of the storage period, was reduced greatly when compared to the use of fresh sperm (Table 1.1). In fact, the lower fertilization rate has been reported to be the major factor in the reduced fertility found after *in vitro* storage[30].

In addition to a greater proportion of unfertilized oocytes, abnormalities of the preimplantation period have been observed following sperm storage. Gulyas[29] found that 11 out of 22 rabbit ova, fertilized by spermatozoa stored for six days, had proceeded through the first cleavage division but subsequently degenerated. Blastomeres had separated and were pycnotic by the time the zygotes were examined 26–27 h after insemination of the female. This was not observed in zygotes resulting from oocytes fertilized by either fresh or three-day-old sperm. When two-cell stages, recovered from female mice inseminated with epididymal sperm aged up to 24 h, were cultured to the blastocyst stage, it was found that embryonic mortality increased significantly following 18 h sperm storage[34].

The chromosome complement of rabbit blastocysts arising from spermatozoa stored *in vitro* at 5 °C has been the subject of two investigations[31, 37]. Basically the same procedure was used in each study. Blastocysts resulting from fresh, two- or six-day-old sperm were examined in one case[37] and, in the other investigation[31], sperm stored at daily intervals up to five days were used to inseminate the females. The results are summarised in Table 1.2. In the study by Nicolai and Shaver[31], a fertility peak was noted when spermatozoa were stored for three days. The recovery rate reached 95% and exceeded the control value of 85%. This increase in fertility was not observed by Stranzinger and Lodge[37]. However, in their study spermatozoa were aged for two and six days only and a fertility peak may have been missed.

The two studies also differ in the frequency and types of chromosome abnormalities. Stranzinger and Lodge[37] found aneuploidy to decrease and mosaicism to increase with increasing sperm storage time. In addition,

Table 1.2 Sperm ageing effect at 5 °C on the chromosome complement of rabbit blastocysts

	Storage time (days)	No. females	Blastocyst recovery (%)	No. blastocysts analysed	Chromosome abnormalities (number)				
					Aneuploid	Triploid	Haploid	Tetraploid	Mosaic
A.	0	4	74.5	36	3	1			5
	2	9	35.6	68	3	3	2		8
	6	8	17.9	37	1			2	8
B.	0	6	85	53		1			1
	1	5	73	38					5
	2	6	52	22		1			2
	3	6	95	49	1				2
	4	5	63	37					1
	5	4	27	16					

A. Data from Stranzinger and Lodge (1974)[37]
B. Data from Nicolai and Shaver (1977)[31]

haploid and tetraploid blastocysts were recovered. In the study by Nicolai and Shaver[31], the greatest frequency of chromosomally abnormal blastocyts appeared following sperm storage for 24 or 48 h, when 13.2 and 13.6% of the blastocysts were abnormal. This was immediately before the fertility peak observed at three days of semen storage and was interpreted to lend further support to Salisbury's hypothesis that abnormal sperm compete effectively with normal sperm for fertilization sites at first but later they are unable to do so. This would account for the increased percent recovery of blastocysts and the decline in the frequency of chromosome abnormalities found at the longer sperm storage times. Of the 10 mosaic blastocysts recovered, six were 2n—1/2n and four were 2n/4n. These chromosome abnormalities could have arisen during the early cleavage divisions through non-disjunction or anaphase lagging in the case of the monosomy/diploid blastocysts and failure of cytokinesis to occur with the diploid/tetraploid zygotes. It is interesting to note, however, that mosaicism was the most frequent type of chromosome abnormality found in both studies following storage of spermatozoa at 5 °C. Also 'clustering' of abnormalities occurred; that is, within certain females more than one abnormal blastocyst was recovered. This phenomenon has been seen previously[38] and may be an example of a genetic or physiological influence by the female over the production of heteroploid zygotes. Obviously, further research is needed on the effect of storing spermatozoa at temperatures above freezing on the chromosome complement of zygotes, both in the rabbit and in other species, before definitive conclusions can be drawn.

The effect of sperm storage at temperatures above freezing on postimplantation survival of the conceptus has been limited in most cases to comparing the litter size at birth of females inseminated with either fresh or stored sperm. Therefore, it is not clear whether a decreased litter size is due to loss occurring before or after implantation. Conflicting results have been reported. One study indicated that postimplantation mortality was increased[35]. Rabbit conceptuses were examined at gestation days 8 and 20 and a significant increase in postimplantation loss was found in females inseminated with sperm aged for 48 h at 5 °C, in which loss was 21.6%, compared with the 10.5% loss in females inseminated with fresh sperm. Others believe that increased preimplantation loss accounts for the fertility decline[30]. In order to clarify this important point, fetuses should be examined during the postimplantation period from a variety of species. The number of implantation sites compared with the number of corpora lutea in the ovaries, the number of resorption sites and any phenotypic anomalies should all be carefully recorded. This is an area that requires investigation in order to assess fully the impact of sperm storage on fertility.

In vitro storage below 0 °C

Conflicting reports have been published concerning the effect on fertility of storing spermatozoa at temperatures below freezing. Spermatozoan survival has been shown to be influenced by extender composition, treatment during the freezing and thawing processes, storage times and temperatures. Also, variation occurs between males of the same species and between ejaculates from the same male. Most of the studies deal with the final outcome of pregnancy, that is whether or not a live offspring is produced. Very few investigators have studied the conceptus during gestation.

The preimplantation stages have been examined most frequently in the rabbit following artificial insemination with spermatozoa stored *in vitro* at $-196\,°C$. A significant decrease in the number of cleaved ova was observed following sperm storage for one day or six months compared with fresh sperm[39]. The first 24 h following freezing appears to be the most critical period of any storage interval up to one year. In a study of the 4–16 cell stage, spermatozoa stored for only 5 h produced fewer zygotes than sperm stored for five days or one year[30]. Also, more one-cell and degenerating embryos were found. The percentages of blastocysts recovered also was the lowest following 24 h of sperm storage[40]. The initial reduction in fertility found within the first 24 h of freezing spermatozoa may reflect a period of acclimatization to storage conditions that is required by the male gamete. When cleaved ova were cultured to the blastocyst stage, no significant difference was found between ova penetrated by spermatozoa that had been stored compared with those fertilized by fresh sperm[41]. It was concluded by the latter investigators that, although spermatozoa stored in the frozen state were less able to initiate development, they caused no detectable effect on early embryonic development of the oocytes that had been fertilized.

Recently, a cytogenetic analysis of blastocysts arising from oocytes fertilized by spermatozoa that had been stored for periods up to 24 weeks at $-196\,°C$ was undertaken by Robson[40]. There was little difference between the number of blastocysts recovered from females inseminated with unstored sperm and those inseminated with spermatozoa stored for the varying periods of time. However, the frequency of blastocysts with the chromosome abnormalities rose with increased duration of sperm storage, particularly at the longer intervals of 16 and 24 weeks (Table 1.3). A total of 22 chromosomally

Table 1.3 Chromosome complement of rabbit blastocysts resulting from sperm stored at $-196\,°C$

Storage time (days)	No. females	Blastocyst recovery	No. blastocysts analysed	Chromosome abnormalities (number)		
				Aneuploidy	Triploidy	Mosaicism
0	7	68.5	36			2
1	7	45.5	18			1
28	10	52.1	24		1	3
56	8	57.9	32		2	2
84	7	58.6	27	1		3
112	3	65.5	19			4
168	4	57.1	10	1		2

abnormal blastocysts was recovered with 20 of these arising from the use of stored sperm. Again mosaicism was the most frequent anomaly encountered. Six of the blastocysts were diploid/tetraploid, seven were diploid/monosomic, and two were diploid/trisomic. The abnormalities which were not mosaic included three triploids, one monosomy and one trisomy. Thus mosaicism, probably occurring during the early cleavage divisions, is the predominant chromosomal error following storage of rabbit spermatozoa *in vitro* at either 5 °C or −196 °C.

The postimplantation conceptus has not been studied to the same extent as the preimplantation zygote and blastocyst. Most of the data has come from artificial insemination programmes in cattle. Embryonic mortality was measured from the proportion of cows that required a subsequent insemination following an interval of about two oestrous cycles or 46 days. A 16% mortality of embryos and fetuses occurred between 30 and 180 days postbreeding, with most of the loss found before 90 days[42]. As techniques for storage of bull spermatozoa at −196 °C have improved, fertility levels have been maintained for longer periods of time. At present, little loss in fertility occurs with storage of bull spermatozoa for periods up to five years[27,42]. Also, a slight improvement in fertility, similar to the fertility peak found after sperm storage at 5 °C, was observed following sperm storage at −196 °C for 3–4 months[42].

Sherman[43] has compiled data from human sperm banks and found no increase in the frequency of abortions or children with birth defects. However, data were compiled from only 1464 cases and the length of time that spermatozoa were stored for each of these inseminations was not recorded. There were 113 spontaneous abortions and 11 abnormal children born. No details were given either of the abnormalities or the chromosome complement of the abortuses.

The effect which the storing of spermatozoa *in vitro* at −196 °C has on development and particularly on the chromosome complement of the conceptus is an area that definitely requires further investigation. Data from a variety of species are needed with careful attention paid to sperm storage conditions and length of storage time.

POSSIBLE MECHANISMS UNDERLYING ABNORMALITIES RESULTING FROM SPERM AGEING

The various studies presented clearly indicate that when spermatozoa are stored for extended periods *in vivo* or *in vitro* their ability to contribute to chromosomally normal embryos is lost before the ability to fertilize ova. The question that arises is whether or not the abnormalities generated are a direct result of genomic alterations in the aged spermatozoa or a result of delayed fertilization, which is egg ageing. Cleavage errors[9], nuclear fragmentation[44], triploidy[8,45] and non-disjunction[7,46] have all been shown to result from egg ageing.

Martin-DeLeon and associates[12], in analysing the types of abnormalities generated by sperm aged in the male tract, argued that all the anomalies observed, namely trisomy, triploidy, mosaicism and the deletion could be attributed to an 'ovum factor' rather than a 'sperm factor' as proposed by Bedford[47]. They suggested that aged sperm might be impaired in their ability to progress to the fertilization site and initiate normal activation of the egg in the required time span. Evidence[48] that embryological anomalies have been due to reduced motility of sperm resulting in delayed fertilization and thus egg senescence has been presented.

In order to accurately define the specific role of the ageing sperm in generating chromosomal aberrations, Thibault[22] undertook a study in the rabbit in which fresh eggs were fertilized *in vitro* by spermatozoa aged *in utero*. This procedure allowed the precise timing of contact between egg and sperm. The results of this investigation clearly demonstrated a significant increase in nuclear abnormalities, which included digyny and abnormal male pronuclear development, in fertilized eggs as a consequence of sperm ageing. In the light of these results and findings of nuclear defects induced by sperm ageing in the fruit-fly[49] and in the frog[50], Salisbury and Hart[26] in a recent review concluded that changes in the sperm genome are involved in the underlying mechanism of embryonic abnormalities resulting from sperm ageing.

Biochemical studies designed to detect changes in the DNA of the sperm nucleus have failed to show alterations in DNA content during sperm ageing *in vivo* or *in vitro*[51]. However, changes in the deoxyribonucleoprotein (DNP) as a whole have been implicated in sperm senescence. These include an increased number of antigens in the DNP extracted from aged bull sperm as compared to fresh ones[52], changes in the thermal stability and hyperchromicity with age[53] and decreased Feulgen stainability of spermatozoa aged both *in vivo* and *in vitro*[54-56] possibly due to a corresponding disulphide formation within and/or among spermatozoan nuclear proteins[57].

Studies employing sodium dodecyl sulphate and dithiothreitol assay, a sensitive measure of disulphide content, have detected an increased disulphide content in *in vivo* aged rabbit sperm[58] and *in vitro* stored bull sperm at 25 °C[59]. The formation of disulphide cross-linkages is increased *in vivo* during nuclear maturation of the spermatozoon in the male reproductive tract[58]. Nuclear decondensation of the sperm nucleus is accomplished normally soon after penetration of the oocyte. With increased disulphide cross-linkages, the decondensation of the sperm nucleus and the formation of the male pronucleus could be affected leading ultimately to an abnormal genomic expression. Thibault[22] has observed an increase in the formation of abnormal male pronuclei when aged rabbit spermatozoa were used during *in vitro* fertilization. Longo and Kunkle[60] suggested that the transformation of condensed sperm chromatin upon entry of the sperm into the egg cytoplasm may allow the paternal genome to participate in nucleic acid synthesis. Failure of this transformation or decondensation in the ageing sperm may lead to a deficiency of the necessary factors required for normal mitosis. Thus, chromosome aberrations generated by the ageing sperm may be explained by changes in the spermatozoan DNP.

SUMMARY

1. Inseminations with spermatozoa subjected to prolonged *in vitro* or *in vivo* storage result in a greater embryonic loss as compared to inseminations with freshly ejaculated spermatozoa.

2. Increased frequencies of gross chromosomal aberrations have been detected in embryos resulting from sperm stored in both environments. The abnormalities detected in six-day rabbit blastocysts when sperm were aged in

the male tract following bilateral ligation of the corpus epididymides included trisomy, triploidy, and mosaicism of the euploid series.

3. The main abnormality seen when spermatozoa were aged *in utero* and *in vitro* was mosaicism both of the aneuploid and euploid series. An interesting finding was sex chromosome chimeras seen in the series of blastocysts resulting from sperm aged *in utero*.

4. Further research is needed in all areas of sperm ageing using different species to determine the frequencies of phenotypic and chromosomal anomalies in early and later stages of gestation.

5. The possible mechanisms underlying the induction of chromosome abnormalities have been discussed. It is probable that chromosome aberrations generated by sperm ageing *in vitro* and *in vivo* may be explained by changes in the disulphide content of the deoxyribonucleoprotein complex.

ACKNOWLEDGEMENTS

We wish to acknowledge the technical assistance of Miss Isobel Morrison. The authors' research was supported by the Medical Research Council of Canada. Support for the first author during the preparation of this manuscript was provided by the UNIDEL Human Heredity Program.

References

1. Salisbury, G. W. and Hart, R. D. (1970). Gamete aging and its consequences. *Biol. Reprod.*, 2 (Suppl.), 1
2. Vander Vliet, W. L. and Hafez, E. S. E. (1974). Survival and aging of spermatozoa: A review. *Am. J. Obstet. Gynecol.*, 118, 1006
3. Salisbury, G. W., Hart, R. G. and Lodge, J. R. (1976). The fertile life of spermatozoa. *Perspect. Biol. Med.*, 19, 213
4. Guerrero, V. and Rojas, O. I. (1975). Spontaneous abortion and aging of human ova and spermatozoa. *N. Engl. J. Med.*, 293, 573
5. Tesh, J. M. and Glover, T. D. (1969). Ageing of rabbit spermatozoa in the male and its effect on fertility. *J. Reprod. Fertil.*, 20, 287
6. Carr, D. H. (1971). Chromosome studies in selected spontaneous abortions. III. Early pregnancy loss. *Obstet. Gynecol.*, 37, 750
7. Witschi, E. and Laguens, R. (1963). Chromosomal aberrations in embryos from overripe eggs. *Dev. Biol.*, 1, 605
8. Shaver, E. L. and Carr, D. H. (1967). Chromosome abnormalities in rabbit blastocysts following delayed fertilization. *J. Reprod. Fertil.*, 14, 415
9. Vickers, A. D. (1969). Delayed fertilization and chromosomal anomalies in mouse embryos. *J. Reprod. Fertil.*, 20, 69
10. Young, W. C. (1931). A study of the function of the epididymis. III. Functional changes undergone by spermatozoa during their passage through the epididymis and vas deferens of the guinea pig. *J. Exp. Biol.*, 8, 15
11. Igboeli, G. and Foote, R. H. (1969). Maturation and aging changes in rabbit spermatozoa isolated by ligatures at different levels of the epididymis. *Fertil. Steril.*, 20, 506
12. Martin-DeLeon, P. A., Shaver, E. L. and Gammal, E. B. (1973). Chromosome abnormalities in rabbit blastocysts resulting from spermatozoa aged in the male tract. *Fertil. Steril.*, 24, 212
13. Tesh, J. M., Walker, S. and Tesh, S. A. (1973). The ageing sperm as a possible cause of abnormal prenatal development. *Teratology*, 8 (2), 240
14. Ekins, J. G. and Shaver, E. L. (1976). Cytogenetics of postimplantation rabbit conceptuses following delayed fertilization. *Teratology*, 13, 57

15. Martin-DeLeon, P. A. (1979). A developmental and karyological study of mouse blastocysts resulting from spermatozoa aged *in vivo*. (In preparation)
16. Nalbandov, A. V. and Card, L. E. (1943). Effect of stale sperm on fertility and hatchability of chicken eggs. *Poultry Sci.*, **22**, 218
17. Dharmarajan, M. (1950). Effect on the embryo of staleness of the sperm at the time of fertilization in the domestic hen. *Nature*, **165**, 398
18. Tesh, J. M. (1969). Effects of the ageing of rabbit spermatozoa *in utero* on fertilisation and prenatal development. *J. Reprod. Fertil.*, **20**, 299
19. O'Ferrall, G. J. M. (1973). Effect of varying the time of artificial insemination in relation to ovulation on conception and prenatal losses in the rabbit. *Biol. Reprod.*, **9**, 338
20. Maurer, R. R., Whitener, R. H. and Foote, R. H. (1969). Relationship of *in vivo* gamete aging and exogenous hormones to early embryo development in rabbits. *Proc. Soc. Exp. Biol. Med.*, **131**, 882
21. Harper, M. J. K. (1970). Cytological observations on sperm penetration of rabbit eggs. *J. Exp. Zool.*, **174**, 141
22. Thibault, C. (1971). Normal and abnormal fertilization in mammals. *Adv. Biosci.*, **6**, 63
23. Martin, P. A. and Shaver, E. L. (1972). Sperm aging *in utero* and chromosomal anomalies in rabbit blastocysts. *Dev. Biol.*, **28**, 480
24. Lodge, J. R., Ax, R. L. and Fechheimer, N. S. (1974). Chromosome aberrations in embryos from *in vivo* aged chicken sperm. *Poultry Sci.*, **53**, 1816
25. Boué, J., Boué, A. and Lazar, P. (1975). The epidemiology of human spontaneous abortions with chromosomal anomalies. In: R. J. Blandau (ed.). *Aging Gametes*, pp. 330–348 (Basel: S. Karger)
26. Salisbury, G. W. and Hart, R. G. (1975). Functional integrity of spermatozoa after storage. *Bioscience*, **25**, 159
27. Foote, R. H. (1978). Maintenance of fertility of spermatozoa at $-196\,°C$. In: *The Integrity of Frozen Spermatozoa*. (Nat. Acad. Sci., Washington, D.C.) pp. 144–162
28. Sherman, J. K. (1973). Synopsis of the use of frozen human semen since 1964: State of the art of human semen banking. *Fertil. Steril.*, **24**, 397
29. Gulyas, B. J. (1968). Effects of aging on fertilizing capacity and morphology of rabbit sperm. *Fertil. Steril.*, **199**, 453
30. Maurer, R. R., Stranzinger, G. F. and Paufler, S. K. (1976). Embryonic development in rabbits after insemination with spermatozoa stored at 37, 5 or $-196\,°C$ for various periods. *J. Reprod. Fertil.*, **48**, 43
31. Nicolai, P. and Shaver, E. L. (1977). The chromosome complement of rabbit blastocysts resulting from spermatozoa stored at $5\,°C$. *Biol. Reprod.*, **17**, 640
32. Salisbury, G. W. and Flerchinger, F. H. (1967). Aging phenomena in spermatozoa. I. Fertility and prenatal losses with use of liquid semen. *J. Dairy Sci.*, **50**, 1675
33. Stranzinger, G. and Paufler, S. (1971). Der Einfluss der *in vitro* Spermaaufbewahrung auf die Befruchtung und den Fruchttod beim Kaninchen. *Zuchthygiene*, **6**, 99
34. Benitz, A. M. and Graves, C. N. (1972). In-vitro aging of mouse spermatozoa. *J. Anim. Sci.*, **35**, 236
35. Koefoed-Johnsen, H. H., Pavlok, A. and Fulka J. (1971). The influence of aging of rabbit spermatozoa *in vitro* on fertilizing capacity and embryonic mortality. *J. Reprod. Fertil.*, **26**, 351
36. First, N. L., Stratman, F. W. and Casida, L. E. (1963). Effect of sperm age on embryo survival in swine. *J. Anim. Sci.*, **22**, 135
37. Stranzinger, G. F. and Lodge, J. R. (1974). Sperm aging effects on the chromosomal complement in rabbit blastocysts. *Z. Tier. Zucht.*, **91**, 125
38. Fechheimer, N. S. and Beatty, R. A. (1974). Chromosomal abnormalities and sex ratio in rabbit blastocysts. *J. Reprod. Fertil.*, **37**, 331
39. O'Shea, T. and Wales, R. G. (1969). Further studies of the deep freezing of rabbit spermatozoa in reconstituted skim milk powder. *Aust. J. Biol. Sci.*, **22**, 709
40. Robson, K. E. (1977). A cytogenetic study of rabbit blastocysts resulting from spermatozoa stored *in vitro* at $-196\,°C$. M.Sc. Thesis, Univ. Western Ontario, London, Canada
41. Maurer, R. R. and Foote, R. H. (1972). Effects of frozen semen and insemination time on early embryonic development in the rabbit. *Biol. Reprod.*, **7**, 103

42. Lee, A. J., Salisbury, G. W., Boyd, L. J. and Ingalls, W. (1977). *In vitro* aging of frozen bull semen. *J. Dairy Sci.*, **60**, 89
43. Sherman, J. K. (1978). Banks for frozen human semen: Current status and prospects. In: *The Integrity of Frozen Spermatozoa*. (Washington Nat. Acad. Sci.), pp. 78–91
44. Beatty, R. A. (1957). *Parthenogenesis and Polyploidy in Mammalian Development* (London: Cambridge University Press)
45. Austin, C. R. and Braden, A. W. H. (1953). An investigation of polyspermy in the rat and rabbit. *Aust. J. Biol. Sci.*, **6**, 674
46. Mikamo, K. (1968). Mechanism of non-disjunction of meiotic chromosomes and of degeneration of maturation spindles in eggs affected by intrafollicular overripeness. *Experientia*, **24**, 75
47. Bedford, J. M. (1966). Development of the fertilizing ability of spermatozoa in the epididymis of the rabbit. *J. Exp. Zool.*, **163**, 319
48. Orgebin-Crist, M. C. (1968). Maturation of spermatozoa in the rabbit epididymis: Delayed fertilization in does inseminated with epididymal spermatozoa. *J. Reprod. Fertil.*, **16**, 29
49. Byers, H. and Muller, H. J. (1952). Influence of aging at two different temperatures on the spontaneous mutation rate of mature spermatozoa of *Drosophila melanogaster*. *Genetics*, **37**, 570
50. Hart, R. G. and Salisbury, G. W. (1967). The effect of sperm age on embryonic mortality in the frog. *Fed. Proc.*, **26**, 645
51. Mann, T. and Lutwak-Mann, C. (1975). Biochemical aspects of aging in spermatozoa in relation to motility and fertilizing ability. In: R. J. Blaudau (ed.). *Aging Gametes*, pp. 122–150 (Basel: S. Karger)
52. Todorovic, R. C., Graves, C. N. and Salisbury, G. W. (1969). Aging phenomena in spermatozoa. IV. Immunoserologic characterization of the deoxyribonucleic acid-protein extracted from fresh and aged bovine spermatozoa. *J. Dairy Sci.*, **52**, 1415
53. Beil, R. E. (1975). Alterations of chromatin during the *in vitro* storage of bovine spermatozoa. Ph.D. Thesis, University of Illinois, Urbana, Illinois
54. Bouters, R. C., Esnault, R. and Salisbury, G. W. (1967). Comparison of DNA revealed by Feulgen and by ultraviolet light in the rabbit spermatozoa after storage in the male efferent ducts. *Nature*, **213**, 181
55. Bouters, R., Esnault, C., Salisbury, G. W. and Ortavant, R. (1967). Discrepancies in analyses of DNA in rabbit spermatozoa involving Feulgen staining (Feulgen-DNA) and ultraviolet light absorption (UV-DNA) measurements. *J. Reprod. Fertil.*, **14**, 355
56. Salisbury, G. W., Birge, W. J., De La Torre, L. and Lodge, J. R. (1961). Decrease in the nuclear Feulgen-positive material (DNA) upon aging in *in vitro* storage of bovine spermatozoa. *J. Biophys. Biochem. Cytol.*, **10**, 353
57. Esnault, C. (1973). Reactivation of the Feulgen reaction of ram spermatozoa by dithiothreitol. *J. Reprod. Fertil.*, **32**, 153
58. Calvin, H. I. and Bedford, J. M. (1971). Formation of disulphide bonds in the nucleus and accessory structures of mammalian spermatozoa during maturation in the epididymis. *J. Reprod. Fertil.*, **13** (Suppl. 1), 65
59. Beil, R. E. and Graves, C. N. (1977). Nuclear decondensation of mammalian spermatozoa: Changes during maturation and *in vitro* storage. *J. Exp. Zool.*, **202**, 235
60. Longo, F. J. and Kunkle, M. (1977). Synthesis of RNA by male pronuclei of fertilized sea urchin eggs. *J. Exp. Zool.*, **201**, 431

2
The influence of teratogens on mouse embryo chromosome structure

R. G. SKALKO, S. M. TUCCI AND R. H. DROPKIN

INTRODUCTION

The rapid and extensive development of newer methods to probe the complexities of chromosome structure has been one of the great achievements of modern biology. The analysis of the many banding patterns observed in metaphase chromosome preparations[1] and the use of specific fluorochromes to evaluate the molecular components of chromatin[2-4] have permitted the description of specific differences between chromosomes in a particular metaphase preparation[5,6] as well as differences within the chromosome itself[7]. Coincident with these technical achievements, much work has been done to use them in the evaluation of actual and potential disease states in humans[8,9]. The following is a report on our attempts to utilize these methods and their extension, the analysis of sister chromatid exchange (SCE)[10], to demonstrate the effect of documented teratogens on chromosome structure.

Interest in the relationship between the ability of a teratogen to produce an embryotoxic response and to affect chromosome structure is not new but prior studies were sporadic in nature due in no small part to the relative crudeness of the methods in use at the time they were performed. Although restricted to an analysis of chromosomal aberrations in the classical sense, they provided a factual and conceptual framework upon which our own efforts were based. Ingalls *et al.*[11] demonstrated that a teratogenic dose of 6-aminonicotinamide (6-AN), administered to pregnant mice on day 13 of gestation, produced a high incidence of cleft palate and a measurable increase in the number of isolated metaphases from the embryo which exhibited polyploidy and fragmentation. This observation was followed by two studies from Warkany's laboratory[12,13] which showed that embryotoxic doses of X-irradiation, in particular, were capable of producing a striking increase in the number of abnormal metaphases in cells isolated from treated rat embryos within the first 24 h after treatment. Less spectacular results were obtained following the administration of teratogenic doses of nitrogen mustard, chlorambucil and streptonigrin[13]. Other studies utilizing standardized methods have also appeared[14,15] but they, too, have been restricted to the analysis of chromosomal aberrations such as anaphase bridges, gaps, breaks and terminal deletions.

METHODS

Virgin albino mice of the ICR strain were used exclusively and the details of breeding and the isolation of pregnant females were identical to those employed in previous studies[16]. Teratogenic and chromosomal analyses were done on embryos whose mothers were treated on day 10 of gestation (day 0 = plug date).

Teratology studies

Two teratogens were analysed in detail. These were: the halogenated pyrimidine nucleoside 5-bromo-2'-deoxyuridine (BrdU)[17] and the antibiotic mitomycin C (MMC)[18]. Both compounds were obtained from Schwarz-Mann and were dissolved under aseptic conditions, BrdU as a 2% solution, MMC as an 0.1% solution. MMC was prepared just prior to use while BrdU was satisfactorily stored at room temperature. Females treated with BrdU received intraperitoneally either 500, 1000 or 2000 mg/kg. The dosages of MMC administered were 5 and 10 mg/kg. Treated females and contemporaneous untreated controls were killed on day 17 and their conceptuses analysed by standard methods[17].

Chromosomal studies

Independent of the specific experimental protocol used, chromosomes were obtained from either day 10 or day 11 embryos in the following manner. Individual females were killed by cervical dislocation and the embryos dissected free of all extraembryonic membranes. A cell suspension was prepared by mild agitation with a pasteur pipette in a sterile medium containing RPMI 1640, 15% fetal calf serum and 0.1 μg/ml colcemid (all supplied by Grand Island Biological Co.). The cells were then cultured in the same medium at 37 °C for 2 h. At the end of the culture period, the cells were concentrated by mild centrifugation, the medium decanted and the pellet resuspended in 0.075 M KCl for 10 min. After further centrifugation, the cells were resuspended in fixative (ethanol:acetic acid; 3:1) and were fixed overnight. Chromosome spreads were made by dropping aliquots of the cell suspension onto precleaned, cold slides.

Three specific staining procedures were employed. In the first, air-dried slides were stained with Giemsa. A stock solution of this stain was prepared by dissolving 1 g of Giemsa R66 (Gurr) in 66 ml of glycerol at 55–60 °C for 1.5–2.0 h. Following this, 66 ml of absolute methanol were added. A fresh working solution was made up daily by making an aliquot of the stock solution to be 2% in a pH 6.8 buffer (Gurr). Slides were immersed in the stain for a determined time period (1.5–8.0 h), rinsed in water, air-dried and finally mounted in DPX (Gurr).

Two methods were employed to demonstrate Giemsa-bands (G-bands)[1]. The first was the acetic/saline/Giemsa (ASG) procedure[19]. In this, slides were exposed to 2 × SSC (0.3 M NaCl and 0.3 M trisodium citrate) for 1 h at 65 °C. They were briefly rinsed in distilled water and then stained for 30 min in Giemsa stain (above). In the second, slides were treated for 1 h at 65 °C in Hank's balanced salt solution (BSS) without calcium and magnesium[20]. Subsequent treatment was identical to that used in the ASG technique.

The demonstration of sister chromatid exchanges (SCEs), using present methods, requires that cells be exposed to BrdU continually during two consecutive cell cycles. This results in quenched fluorescence of the fluorochrome 33258 Hoechst in bifilarly substituted chromatids[10,21]. This phenomenon is also observed if fluorochromed preparations are subsequently stained with Giemsa (fluorescence plus Giemsa or FPG)[22]. Since a single, biologically effective, dose of BrdU (e.g. 500 mg/kg) results in only a transient exposure to the embryo *in vivo*[23], we employed a modification of the procedure of Allen and Latt[24] to insure chronic exposure *in utero* over two cell cycles. Females were administered 15 consecutive hourly injections, intraperitoneally, of 25 mg/kg BrdU. Parallel experiments showed that this procedure was not teratogenic using standard procedures[17]. Following the final injection, metaphase spreads were prepared as described above and slides were stained for 10–15 min in 33258 Hoechst[24]. The slides were washed, mounted in distilled water and analysed. Some preparations were also stained with Giemsa (FPG)[22]. BrdU was not used as a teratogen for these experiments since its use was an essential component of the technique itself. However, to study the effect of MMC, a range of dosages (0.5–10.0 mg/kg) was administered, intraperitoneally, 30 min before the final four BrdU injections.

All preparations were observed and photographed with a Zeiss Photomicroscope II. Giemsa-stained preparations were photographed with Panatomic-X film (Kodak). For fluorescence, a BG 12 excitation filter and BG 53 and 44 barrier filters were used and the preparations photographed with Tri-X film (Kodak).

RESULTS

Teratology

Treatment of pregnant females with BrdU resulted in a dose-dependent increase in all parameters for embryotoxicity that were measured (Table 2.1), most obviously in the occurrence of the digital reduction defects, ectrodactyly and syndactyly. Similar results were obtained after MMC treatment (Table 2.2). The occurrence of spontaneous cleft palate in this experiment has been reported once before for this strain[25] and this latest observation emphasizes once again the necessity to use contemporaneous controls in all testing procedures. Our results are quantitatively different from those obatined by Tanimura[18] and suggest that with MMC toxicity, also, there may be distinct strain differences.

Chromosome analysis

With a few exceptions, all treated females were administered dosages of either BrdU or MMC on the morning of day 10 of gestation. They were killed 4–5 h later and chromosome spreads prepared. For BrdU-treated embryos, we observed altered Giemsa-staining patterns, the occurrence of chromosome breaks in metaphase spreads and the occurrence of unequal banding of sister chromatids (lateral asymmetry) after G-banding[20]. For MMC-treated embyros, the occurrence of dose-dependent increases in chromosome breaks[26] was studied, independently, by two of us (S. M. T., R. H. D.). In addition, preparations were analysed for lateral asymmetries and SCEs.

BrdU

A basic observation related to BrdU incorporation into chromosomal DNA[23] has been that its substitution for thymidine (dT) results in a decrease in Giemsa staining intensity in highly substituted regions of the chromosome[21, 27, 28]. It has been shown that a bias exists in the distribution of AT pairs in the chromosome, with satellite DNA[29] having a preponderance of AT sites. Since there is also evidence that this fraction is restricted to the centromeres of mouse chromosomes[30], we treated pregnant mice with 2000 mg/kg BrdU (Table 2.1) and made chromosome preparations 2 h later. Using this protocol, only cells in late S would incorporate BrdU into their DNA. The lighter staining centromeres (Figure 2.1) confirm the observation that

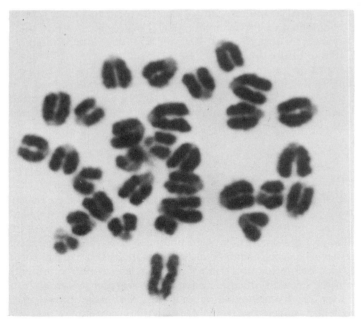

Figure 2.1 Giemsa-stained chromosomes from a 10-day mouse embryo exposed *in utero* to 2000 mg/kg BrdU for 2 h. Centromeric regions (satellite DNA) are lightly stained

Table 2.1 Embryotoxic effects of BrdU on the 10-day mouse embryo

Treatment	No. females	No. of sites	Resorbed		Survivors	Survivors with					
						CP		E		S	
			No.	%		No.	%	No.	%	No.	%
Control	11	135	8	(5.9)	127	—	—	—	—	—	—
500 mg/kg	11	151	10	(6.6)	141	95	(67.4)	1	(0.7)	2	(1.4)
1000 mg/kg	11	122	24	(19.7)	98	95	(96.9)	49	(50.0)	26	(26.5)
2000 mg/kg	13	187	90	(48.1)	97	96	(99.9)	81	(83.5)	38	(39.2)

Abbreviations used: CP = cleft palate; E = ectrodactyly; S = syndactyly

Table 2.2 Embryotoxic effects of MMC on the 10-day mouse embryo

Treatment	No. females	No. of sites	Resorbed		Survivors	Survivors with					
						CP		E		S	
			No.	%		No.	%	No.	%	No.	%
Control	12	145	5	(3.5)	140	4	(2.9)	—	—	—	—
5 mg/kg	11	149	10	(6.7)	139	18	(12.9)	49	(41.7)	22	(15.8)
10 mg/kg	11	148	52	(35.1)	96	29	(30.2)	77	(80.2)	37	(38.5)

Abbreviations used: CP = cleft palate; E = ectrodactyly; S = syndactyly

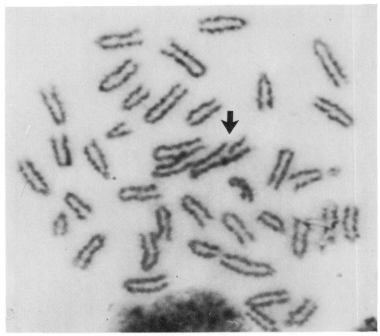

Figure 2.2 Giemsa-stained chromosomes from a 10-day mouse embryo exposed *in utero* to 2000 mg/kg BrdU for 4 h. Arrow indicates a chromosome break

centromeric DNA is synthesized during late S phase[31] and that BrdU can be visualized in the DNA of the embryo by yet another procedure after *in vivo* treatment[23].

When metaphase spreads were analysed for the presence of chromosomal breaks (Figure 2.2) and for an increase in the occurrence of lateral asymmetry in G-banded preparations[20], a dose-dependent relationship in the embryotoxic range was observed (Table 2.3). Comparison of these two parameters of presumed chromosomal damage indicates that the latter is the more sensitive.

Table 2.3 Chromosomal alterations produced by embryotoxic doses of BrdU

Treatment	No. of cells	No. of chromosomes	No. of breaks	Breaks/ cell	No. of cells	No. of chromosomes	LA*	LA/c
0	50	2003	4	0.08	50	2004	89	1.7
500 mg/kg	50	1982	6	0.12	50	1972	170	3.4
1000 mg/kg	50	1878	9	0.18	50	1998	255	5.1
1500 mg/kg	50	2018	14	0.28	50	2009	410	8.2
2000 mg/kg	50	2003	24	0.48	50	2018	560	11.2

* LA = lateral asymmetries in G-banded preparations

Giemsa staining[47], this effect appears to be non-specific[50]. It has been suggested however, that the presence of some of the histone components are essential for banding[47,51] and that DNA : non-histone interactions may also be involved[48,52]. Thus, the interpretation of the results reported here and elsewhere[24,32,34] ultimately depends on the acquisition of more information on the basic organization of chromosomes and how the various staining techniques represent this organization[52].

ACKNOWLEDGEMENTS

We thank two eager medical students, Ms Diane Traver and Ms Ginger Miller who, as Biology Honours students at Russell Sage College, spent many hours in helping us develop the techniques used in this study. They were the true pioneers in probing the intricacies of embryo chromosome structure and we are for ever in their debt. One of us (R. H. D.) carried on their efforts as a student in the Biomedical Program administered jointly by Rensselear Polytechnic Institute and Albany Medical College. We express our deep appreciation to Dr Ian H. Porter, Director of the Birth Defects Institute, for his continual support and encouragement of our work. Our thanks also go to Allen M. Niles, Lin S. Smith, Louise L. Skalko, and Liz DeLoach for their invaluable assistance at critical junctures in this study. Partial support was provided by a research grant to R. G. S. from the National Foundation-March of Dimes.

References

1. Schnedl, W. (1974). Banding patterns in chromosomes. *Int. Rev. Cytol.*, **Suppl. 4**, 237
2. Caspersson, T., Lindsten, J., Lomakka, E., Moller, A. and Zech. L. (1972). The use of fluorescence techniques for the recognition of mammalian chromosomes and chromosome regions. *Int. Rev. Exp. Pathol.*, 11, 1
3. Hsu, T. C. (1973). Longitudinal differentiation of chromosomes. *Annu. Rev. Genet.*, 7, 153
4. Moutschen, J. (1976). Fine structure of chromosomes as revealed by fluorescence analysis. *Prog. Biophys. Mol. Biol.*, 31, 39
5. Drets, M. E. and Shaw, M. W. (1971). Specific banding patterns of human chromosomes. *Proc. Natl. Acad. Sci. (USA)*, 68, 2073
6. Miller, D. A. and Miller, O. J. (1972). Chromosome mapping in the mouse. *Science*, 178, 949
7. Chen, T. R. and Ruddle, F. H. (1971). Karyotype analysis utilizing differentially stained constitutive heterochromatin of human and murine chromosomes. *Chromosoma*, 34, 51
8. Bergsma, D. and Schimke, R. N. (1976). *Cytogenetics, Environment and Malformation Syndromes*. (New York: Alan R. Liss)
9. Hook, E. B. and Porter, I. H. (1977). *Population Cytogenetics*. (New York: Academic Press)
10. Wolff, S. (1977). Sister chromatid exchange. *Annu. Rev. Genet.*, 11, 183
11. Ingalls, T. H., Ingenito, E. F. and Curley, F. J. (1963). Acquired chromosomal anomalies induced in mice by injection of a teratogen in pregnancy. *Science*, 141, 810
12. Soukup, S. W., Takacs, E. and Warkany, J. (1965). Chromosomal changes in rat embryos following X-irradiation. *Cytogenetics*, 4, 130
13. Soukup, S., Takacs, E. and Warkany, J. (1967). Chromosomal changes in embryos treated with various teratogens. *J. Embryol. Exp. Morphol.*, 18, 215
14. Joneja, M. and Ungthavorn, S. (1968). Chromosome aberrations in trypan blue induced teratogenesis in mice. *Canad. J. Genet. Cytol.*, 10, 91
15. Bell, L. J., Branstrator, M., Roux, C. and Hurley, L. S. (1975). Chromosomal abnor-

malities in maternal and fetal tissues of magnesium- or zinc-deficient rats. *Teratology*, **12**, 221
16. Skalko, R. G. and Packard, D. S., Jr. (1974). The teratogenic response of the mouse embryo to 5-iododeoxyuridine. *Experientia*, **29**, 198
17. Skalko, R. G., Packard, D. S., Jr., Schwendimann, R. N. and Raggio, J. F. (1971). The teratogenic response of mouse embryos to 5-bromodeoxyuridine. *Teratology*, **4**, 87
18. Tamimura, T. (1968). Effects of Mitomycin C administered at various stages of pregnancy upon mouse fetuses. *Okajimas Fol. Anat. Jap.*, **44**, 337
19. Summer, A. T., Evans, H. J. and Buckland, R. A. (1971). New techniques for distinguishing between human chromosomes. *Nature New Biol.*, **232**, 31
20. Tucci, S. M. and Skalko, R. G. (1977). Demonstration of lateral asymmetry in G-banded mouse embryo chromosomes. *Experientia*, **33**, 1437
21. Latt, S. A. (1976). Analysis of human chromosome structure, replication and repair using BrdU-33258 Hoechst techniques. *J. Reprod. Med.*, **17**, 41
22. Perry, P. and Wolff, S. (1974). New Giemsa method for the differential staining of sister chromatids. *Nature*, **251**, 156
23. Packard, D. S., Jr., Menzies, R. A. and Skalko, R. G. (1973). Incorporation of thymidine and its analogue, bromodeoxyuridine, into embryos and maternal tissues of the mouse. *Differentiation*, **1**, 397
24. Allen, J. W. and Latt, S. A. (1976). Analysis of sister chromatid exchange *in vivo* in mouse spermatogonia as a new test system for environmental mutagens. *Nature*, **260**, 449
25. Skalko, R. G. and Gold, M. P. (1974). Teratogenicity of methotrexate in mice. *Teratology*, **9**, 159
26. Nowell, P. C. (1964). Mitotic inhibition and chromosome damage by mitomycin in human leucocyte cultures. *Exp. Cell Res.*, **33**, 445
27. Zakharov, A. F. and Egolina, N. A. (1972). Differential spiralization along mammalian mitotic chromosomes. I. BrdU-revealed differentiation in Chinese hamster chromosomes. *Chromosoma*, **38**, 341
28. Kihlman, F. A. and Kronberg, D. (1975). Sister chromatid exchange in *Vicia faba*. I. Demonstration by modified fluorescence plus Giemsa (FPG) technique. *Chromosoma*, **51**, 1
29. Flamm, W. G., McCallum, M. and Walker, P. M. B. (1967). The isolation of complementary strands from a mouse DNA fraction. *Proc. Natl. Acad. Sci. (USA)*, **57**, 1729
30. Pardue, M. L. and Gall, J. G. (1970). Chromosomal localization of mouse satellite DNA. *Science*, **168**, 1356
31. Miller, O. J. (1976). Is the centromeric heterochromatin of *Mus musculus* late replicating? *Chromosoma*, **55**, 165
32. Latt, S. A. (1974). Sister chromatid exchanges, indices of chromosome damage and repair: detection by fluorescence and induction by mitomycin C. *Proc. Natl. Acad. Sci. (USA)*, **71**, 3162
33. Perry, P. and Evans, H. J. (1975). Cytological detection of mutagen–carcinogen exposure by sister chromatid exchange. *Nature*, **258**, 121
34. Kato, H. and Shimada, H. (1975). Sister chromatid exchanges induced by mitomycin C: a new method of detecting DNA damage at chromosomal level. *Mutat. Res.*, **28**, 459
35. Kram, D. and Schneider, E. L. (1978). Reduced frequencies of mitomycin C induced sister chromatid exchanges in AKR mice. *Hum. Genet.*, **41**, 45
36. Vogel, W. and Bauknecht, T. (1976). Differential chromatid staining by *in vivo* treatment as a mutagenicity test system. *Nature*, **260**, 448
37. Solomon, E. and Bobrow, M. (1975). Sister chromatid exchange – a sensitive assay of agents damaging human chromosomes. *Mutat. Res.*, **30**, 273
38. Rutter, W. J., Pictet, R. L., Githens, S. and Gordon, J. S. (1975). The mode of action of the thymidine analogue, 5-bromodeoxyuridine, a model teratogenic agent. In: D. Neubert and H. J. Merker (eds.). *New Approaches to the Evaluation of Abnormal Embryonic Development*, pp. 804–821. (Stuttgart: Georg Thieme)
39. Vig, B. K. (1977). Genetic toxicology of mitomycin C, actinomycins, daunomycin and adriamycin. *Mutat. Res.*, **49**, 189
40. Hsu, T. C. and Somers, C. E. (1961). Effect of 5-bromodeoxyuridine on mammalian chromosomes. *Proc. Natl. Acad. Sci. (USA)*, **47**, 396

41. Lin, M. S., Latt, S. A. and Davidson, R. L. (1974). Microfluorometric detection of asymmetry in the centromeric region of mouse chromosomes. *Exp. Cell Res.*, **86**, 392
42. Angell, R. R. and Jacobs, P. A. (1975). Lateral asymmetry in human constitutive heterochromatin. *Chromosoma*, **51**, 301
43. Galloway, S. M. and Evans, H. J. (1975). Asymmetrical C-bands and satellite DNA in man. *Exp. Cell Res.*, **94**, 454
44. Bostock, C. J. and Christie, S. (1976). Analysis of the frequency of sister chromatid exchange in different regions of chromosomes of the kangaroo rat. *Chromosoma*, **56**, 275
45. Smith, G. P. (1976). Evolution of repeated DNA sequences by unequal crossover. *Science*, **191**, 528
46. Dick, C. and Johns, E. W. (1968). The effect of two acetic acid containing fixatives on the histone content of calf thymus deoxyribonucleoprotein and calf thymus tissue. *Exp. Cell Res.*, **51**, 626
47. Retief, A. E. and Ruchel, R. (1977). Histones removed by fixation. Their role in the mechanism of chromosomal banding. *Exp. Cell Res.*, **106**, 233
48. Comings, D. E., Avelino, E., Okada, T. A. and Wyandt, H. E. (1973). The mechanism of C- and G-banding of chromosomes. *Exp. Cell Res.*, **77**, 469
49. Pathak, S. and Arrighi, F. E. (1973). Loss of DNA following C-banding procedures. *Cytogenet. Cell Genet.*, **12**, 414
50. Holmquist, E. P. and Comings, D. E. (1976). Histones and G-banding of chromosomes. *Science*, **193**, 599
51. Meisner, L. F., Chuprevich, T. W. and Inhorn, S. L. (1973). Giemsa banding specificity. *Nature New Biol.*, **245**, 145
52. Comings, D. E. (1975). Chromosome banding. *J. Histochem. Cytochem.*, **23**, 461

3
Interference with steps in collagen synthesis as a biochemical mechanism of teratogenesis

R. S. BHATNAGAR AND R. S. RAPAKA

INTRODUCTION

Interference with collagen synthesis and deposition can be expected to play a significant role in teratogenesis, not only because collagen is the most abundant protein in the body, being the major protein constituent of most tissues and organs of mesenchymal origin, but especially because it is involved in important aspects of cell and tissue differentiation including epithelio-mesenchymal interaction, cell adhesion and alignment and cell-cell communication and information transfer. Collagen has unique physicochemical features which contribute to its many functions. These features are related to its unusual primary structure and its unique conformation. The potential of collagen to fulfil its many roles is not fully achieved until a series of post-translational modifications of newly synthesized collagen chains occur, resulting in the attainment of its triple-helical conformation and the subsequent assembly and stabilization of the molecules into aggregates. Steps in the synthesis of collagen are susceptible to many exogenous influences. Our thesis, reviewed in this chapter, is that interference with steps in collagen biosynthesis is a major biochemical mechanism of teratogenesis.

ROLE OF COLLAGEN IN TISSUE DIFFERENTIATION AND DEVELOPMENT

Collagen is present in the earliest stages of embryonic development and, throughout development and growth of the organism, collagen continues to undergo turnover at significant rates, at first to accommodate the morphogenetic changes accompanying tissue remodelling and growth and, in the fully grown adult, as part of normal tissue attrition and renewal. Its involvement in many aspects of differentiation, morphogenesis and growth is well documented.

The significance of collagen in epithelio-mesenchymal interactions was first recognized by Grobstein[1-3]. The presence of collagen at the interface between the epithelium and the mesenchyme has been confirmed[4,5] and the removal of collagen at the interface by collagenase has been shown to abolish morphogenetic interactions[6]. Collagen involved in such interactions is synthesized by cells of mesenchymal[7,8] as well as of epithelial[9,10] origin. Epithelial collagen may act as a substratum on which cells align during morphogenesis[9,11-14]. Connective tissue cells are known to migrate in a directed manner and are frequently seen adjacent to pre-existing collagen fibres during morphogenesis. The adhesion of cells to collagenous matrices is well known. Collagen substrata are used to grow cells in culture and collagen has been shown to be involved in the aggregation of platelets. Interaction of cells with collagen may play a role in spatial pattern formation. Because collagen occurs in highly organized fibrous networks under physiological conditions, it may direct the course of morphogenesis by steering cells into appropriate geometric and appositional patterns. Reddi[15,16] demonstrated the potential of the collagenous matrix of bone to alter gene expression in responding fibroblasts and suggested that the geometric and surface charge characteristics of the matrix may play a role in such interactions. Although the mechanisms involved in cell attachment to collagen are not fully understood, recent studies[17] suggest that a protein meshwork containing fibronectin and collagen associated with the fibroblast surface may play a role in the binding of the cell to the extracellular matrix, with collagen–collagen interactions contributing significantly to the complex. Collagen and collagenous matrices have been shown to promote differentiation and morphogenesis in a variety of systems including *in vitro* chondrogenesis[18], bone[15,16] and eye[19,20]. It is clear that collagen synthesis and extracellular aggregation must proceed in an orderly manner for morphogenesis to occur normally.

CHEMICAL BIOLOGY OF COLLAGEN

It is now recognized that collagen is not a single protein but rather, a group of closely related but genetically distinct proteins. The most abundant and widely distributed collagen, Type I collagen accounts for two of the known chain types, $\alpha 1(I)$ and $\alpha 2$. The triple helix of Type I collagen contains two $\alpha 1(I)$ chains and one $\alpha 2$ chain. Type I collagen is nearly ubiquitous in the body, being sparse only in cartilage and in basement membranes. Cartilage contains a distinct type of collagen, Type II collagen whose triple helical molecule is made up of three identical chains, $\alpha 1(II)$. Another type of collagen present in many different tissues, especially those supporting an endothelium such as blood vessels and skin, is Type III collagen, made up of three $\alpha 1(III)$ chains. Type III collagen invariably occurs in association with Type I collagen and it is apparently a component of reticulin fibres. Type III collagen may play a role in morphogenesis and development since it occurs in larger proportions during fetal and early developmental stages but is decreased in adult tissues. Various basement membranes are comprised of a unique collagen species,

ROLE OF COLLAGEN IN TERATOGENESIS

Type IV, whose triple helix is also made up of three identical chains, $\alpha 1(IV)$. Basement membrane collagens are present at the interface between epithelia and mesenchymal structures and they may play a direct role in morphogenesis.

The various collagen species are characterized by interaction properties which facilitate their aggregation in modes most suited to their function. All collagens are present in large aggregates in the physiological milieu; such aggregates may contain other structural macromolecules such as elastin or proteoglycans and they may serve as templates for the deposition of mineral. The interaction properties, which facilitate the aggregation and other characteristic features of collagen in relation to its function, are derived from the unusual amino acid composition and a unique conformation. The triple helical conformation imparts the collagen molecule with a high degree of rigidity and ensures the survival of collagen against lysis by non-specific proteases. Collagen almost uniquely contains hydroxyproline and hydroxylysine in significant amounts. Hydroxyproline stabilizes the collagen triple helix. Depending on the tissue and the type of collagen, many of the hydroxylysine residues may be glycosylated, bound in o-glycosidic linkage to a galactose residue which in turn is bound to a glucose residue. The disaccharide appendages regulate fibrillogenesis and participate in interactions of collagen with other macromolecules.

Because of the presence of the unusual, modified residues, the synthesis of collagen involves several steps (Table 3.1). The initially synthesized poly-

Table 3.1 Steps in synthesis of collagen

Step	Chemical modification	Enzyme
1. Transcription	Synthesis of informational macromolecules	*
2. Translation	Synthesis of polypeptides (procollagen)	*
3. Hydroxylation	Synthesis of hydroxyproline and hydroxylysine	Prolyl hydroxylase, lysyl hydroxylase
4. Glycosylation	Formation of glucosyl-galactosyl hydroxylysine	Galactosyl transferase, glucosyl transferase
	↑ Intracellular Steps	
5. Secretion		†
	↓ Extracellular Steps	
6. Proteolysis	Processing of procollagen to collagen	Procollagen peptidase
7. Fibrillogenesis		‡
8. Oxidative deamination	Formation of lysinaldehyde	Lysyl oxidase
9. Cross-linking	Formation of Schiff base or aldol type cross-links	Non-enzymatic reactions

* Transcription and translation are complex biochemical processes involving multiple enzymatic steps
† The secretion of collagen through the cell membrane is poorly understood. Agents which disrupt microtubules, such as cytochalasin and tunicamycin, inhibit secretion
‡ Fibrillogenesis results from interactions between collagen molecules and between collagen and other extracellular molecules

peptide chains contain no hydroxyproline or hydroxylysine. These are synthesized by the hydroxylation of specific proline and lysine residues already incorporated into the polypeptide, by the enzymes prolyl hydroxylase and lysyl hydroxylase respectively. The two hydroxylases catalyse the hydroxylation of their respective substrates in almost identical reactions, requiring molecular oxygen and α-ketoglutarate as cosubstrates and using Fe^{2+} as the prosthetic metal in the catalytic centre. Both reactions require the presence of a reducing agent, the most effective being ascorbic acid. The glycosylation of collagen occurs while the chains are associated with the endoplasmic reticulum. The initial step involving the binding of galactose is catalysed by a specific galactosyl transferase, and the subsequent coupling of glucose is catalysed by glucosyl transferase. Both enzymes require Mn^{2+} for their activity and utilize UDP-sugars as substrates for glycosylation. Manganese-deficiency syndromes may involve this step in collagen synthesis. The glycosylated collagen is secreted from the cell by cytochalasin and vinblastine sensitive mechanisms. Initially synthesized collagen chains are considerably larger than the functional collagen molecules of the extracellular matrix and, in this state, collagen is called procollagen. The extension peptides facilitate the interaction and alignment of three procollagen chains in register and stabilize the assembly through disulphide bonds. Folding of the three chains into the triple helix is completed before the molecule is secreted. Inhibition of proline hydroxylation by certain agents destabilizes the triple helix and thus hinders collagen secretion. The extension peptides of procollagen are rapidly split off in a stepwise manner, soon after secretion into the extracellular matrix, by a specific procollagen peptidase. Collagen fully processed in this manner is able to aggregate into fibres. The cross-linking between collagen chains in the fibre is initiated by the action of lysyl oxidase which oxidatively deaminates specific lysyl residues to an aldehyde form (α-amino adipic-δ-hydroxy, δ-semi aldehyde). The aldehyde residues are highly reactive and react spontaneously with reactive groups on adjacent chains to form cross-links. Lysyl oxidase requires Cu^{2+} for its activity and Cu^{2+} deficiency or chelating agents affect its activity. Cross-linking is also inhibited by agents which bind to the reactive carbonyl groups formed by the oxidative deamination of collagen. Biosynthesis of collagen has been extensively reviewed in several recent publications[21,22].

Regulatory controls govern the synthesis of collagen at all levels of synthesis and during post-translational processing. As indicated in the following sections, defects in collagen can be induced at many levels of its synthesis. These defects contribute to abnormal morphogenesis and development. The synthesis of collagen may be altered at the genetic level, resulting in the expression of genes for collagen chains not normally resident in the affected tissue at a particular stage in morphogenesis. Regulatory controls may be affected resulting in the increased or decreased synthesis of collagen. Defects in collagen may be induced during post-translational processing as a result of interference in enzymatic steps by drugs, heavy metals and other substances. All of those processes can induce defects in morphogenesis.

INTERFERENCE WITH COLLAGEN SYNTHESIS IN TERATOGENESIS

Because the chain phenotype and physicochemical and interaction characteristics of collagen play a significant role in the regulation of morphogenesis and development, any alteration in these characteristics can be expected to contribute to teratogenesis. It is not surprising, therefore, that many agents and conditions which are known to induce teratogenesis also affect collagen synthesis. Table 3.2 summarizes the molecular mechanisms involved in the induction of teratogenesis by such agents. Although we have included genetic level events, such as transcription in Table 3.2, these will not be discussed further since these effects may involve mechanisms not limited to collagen synthesis and post-translational processing. We have, however, included many

Table 3.2 Biochemical mechanisms in teratogenesis by agents which interfere with collagen synthesis

Nature of teratogen	Biochemical mechanism in relation to collagen	Effect on collagen and on tissues
1. Agents which alter gene expression, e.g. viruses and cellular injury	Synthesis of collagen chains not normally resident in the tissue	Alterations in tissue structure and mechanical properties
2. Inhibitors of DNA, RNA and protein synthesis	Generalized inhibition of cellular processes	
3. Heavy metals and chelating agents	(a) Inhibition of prolyl and lysyl hydroxylase	(a) Lowered denaturation temperature and increased degradation of collagen
	(b) Inhibition of lysyl oxidase	(b) Increased solubility of collagen and decreased tensile strength. Tissue fragility
	(c) Inhibition of collagen glycosylation	(c) Defects in aggregation of collagen, resulting in altered mechanical characteristics of tissue and altered structure
4. Steroids	(a) Generalized depression of macromolecular metabolism	
	(b) Decrease in proline hydroxylation	(b) Same as in 3 (a)
5. DON	Stimulation of collagen synthesis concomitant with decreased proteoglycan	Altered proportions of matrix components
6. Metabolic Inhibitors (a) 6 AN, 3 Ac Pyr. (b) Iodoacetate	(a) Decrease in collagen synthesis(?) (b) Increased collagen synthesis	(a) and (b) Altered proportions of matrix components

The mechanisms involved in the biochemical interactions of the various teratogens listed above are discussed in the text wherever necessary

agents whose effect on the properties of collagen may be correlated directly with teratogenic mechanisms although they may also affect biochemical systems other than those involved in collagen synthesis.

Changes in collagen phenotype

Although the nature of collagen chains involved in morphogenetic interactions has not been fully elucidated, because of the inherent differences in the physicochemical and interaction properties of different collagens, such events must involve a high degree of specificity of collagen chains in order that morphogenesis may proceed in a normal, orderly manner. This is borne out by the fact that changes in collagen phenotype are induced by a variety of agents and conditions, which are also of significance in teratogenesis.

The thymine analogue, 5-bromodeoxyuridine (BUdR), interferes with differentiation in a large number of cell types[23-26] and is a potent teratogen[27]. These effects of BUdR may be related to the alteration of collagen phenotype in cells exposed to this drug. When chick chondrocytes are cultured in the presence of BUdR, they stop synthesizing cartilage-specific Type II collagen and, instead, synthesize Type I collagen and trimers of $\alpha 1(I)$[28]. Chick embryo fibroblasts, which usually synthesize normal Type I collagen characterized by the chain composition $[\alpha 1(I)]_2 \alpha 2$, synthesize the α-1 trimer $[\alpha 1(I)]_3$ when grown in the presence of BUdR[29].

The expression of collagen chain types is also altered in cells infected with viruses and in malignant tissues[30-35]. It may be speculated that similar mechanisms may be involved in the aetiology of virus-induced teratogenesis[36-45].

Expression of collagen phenotypes is altered in many types of chemical injury to cells. Recent studies in our laboratory showed that the rates of syntheses of Type III and Type I collagen are dramatically altered in organ cultures of neonatal rat lungs exposed to hyperoxic environments[46] or paraquat[47]. Hyperoxia is known to induce extensive teratogenesis[48], as is paraquat[49]. The molecular mechanisms in both cases may involve cellular injury induced by the highly reactive oxygen free radical superoxide[50, 51]. Previous studies from our laboratory have shown that superoxide anions play a role in the regulation of collagen synthesis[52].

Altered rates of collagen synthesis

The mechanisms involved in regulating the collagen content of tissues are not fully understood. The turnover rate of collagen is high during stages of development, to accommodate the growth and remodelling of tissues. Any increase or decrease in the rate of collagen synthesis would affect differentiative events and contribute to structural malformations if such changes accompanied developmental stages.

Interference with collagen turnover may be a part of the teratogenic action of various steroids[53-58]. Steroids have been claimed to inhibit collagen synthesis[59, 60] and to promote its degradation[61]. It is more likely, however, that the effect of anti-inflammatory steroids may be due to a generalized inhibition of protein synthesis rather than a specific inhibition of collagen synthesis[62].

Viral infection of cells markedly decreases collagen synthesis[63–65]. The inhibition of collagen synthesis in Rous Sarcoma Virus infected chick embryo fibroblasts has been attributed to decreased levels of translatable collagen mRNA[66]. Although different viruses may interact in different ways with cells' genetic apparatus, it may be speculated that disturbances in collagen synthesis play a role in virus-induced teratogenesis.

Interference with regulatory controls of collagen also appears to be a part of the teratogenic effect of 6-diazo-5-oxo-L-norleucine (DON). DON, a glutamine analogue, has been known to be a teratogen for over two decades[67–69]. DON produces a variety of skeletal defects and cleft palate. It has been used as an experimental tool in investigations on abnormal development[70–73]. Our studies showed that DON specifically stimulated collagen synthesis in tibiae from 10-day-old embryonic chicks[74]. The stimulation was insensitive to actinomycin D, suggesting that DON may have interacted with a post-translational regulatory site. DON is a glutamine antagonist and it inhibits the synthesis of sulphated proteoglycans[75], glycoproteins[76] and nucleic acids[77] by inhibiting the synthesis of glucosamine[78]. The inhibition of proteoglycans and glycoproteins has been cited as contributing to the teratogenic action of DON[73,76,79]. Previous studies by the author on the biochemical effects of DON had shown that the syntheses of collagen and of sulphated proteoglycans proceed independently in chick embryo cartilage explants[80]. In order to ascertain the relative contributions of proteoglycan and glycoprotein syntheses and the synthesis of collagen in DON-mediated biochemical disturbances, chick cartilage explants were treated with DON in the presence of glutamine or glucosamine[81]. Under those conditions there was no inhibition of synthesis of glucosamine-derived macromolecules. However, the stimulation of collagen synthesis continued unabated, indicating that the action of DON on collagen synthesis is specific and not related to the inhibition by DON, of other macromolecular syntheses. Our studies point to a specific role for collagen in the induction of teratogenic changes by DON. The teratogenic effect of DON may involve interference with the synthesis of two major classes of structural macromolecules, collagen and protein–polysaccharides, as well as of cell surface glycoproteins.

Several lines of evidence suggest that teratogenesis resulting from defects in energy metabolism, whether hereditary or experimentally induced, may also involve disturbances in collagen synthesis. Collagen synthesis in most tissues is promoted by conditions favouring oxidative metabolism. This is seen in healing wounds, atherosclerotic plaques and other pathological tissues. The osteogenic transformation of cartilage presents the most clearcut evidence for the role of oxidative metabolism in regulating collagen synthesis. The extracellular matrix of cartilage consists of approximately equal amounts of collagen and proteoglycans, that of bone being mostly collagen. Thus conditions favouring osteogenesis may be expected to invoke a major shift in the production of matrix macromolecules, with a greatly increased emphasis on collagen synthesis. Availability of ample amounts of oxygen promotes osteogenesis, whilst lowered oxygen tensions promote chondrogenesis[82]. Since tissues adapt to the metabolic environment by adjusting the levels of enzymes, a very interesting coefficient has been suggested[83] based on the levels of lactate dehydrogenase (LDH), an enzyme correlating with glycolytic metabolism, and malate dehydrogenase (MDH), a key enzyme in the tricarboxylic acid cycle. The quotient LDH/MDH indicates the relative proportions of glycolytic and oxidative energy metabolism in a tissue. This quotient is nearly 2.0 for cartilage, supporting the concept that hypoxic, glycolytic metabolism favours proteoglycan synthesis at the cost of collagen synthesis. In bone the LDH/MDH quotient drops to 0.5, consistent with much greater reliance on oxidative energy metabolism and higher rates of collagen synthesis than those prevailing in cartilage. In metabolic terms this difference may be related to the fact that the primary precursor for proteoglycans, glucosamine-6-phosphate is derived from fructose-6-phosphate, which may accumulate during glycolysis and which is tapped for proteoglycan synthesis. In contrast, collagen synthesis requires a large amino acid pool and consumes much more energy, both of which can be supplied optimally only under conditions where the cells are particularly enriched as a result of greater contribution from oxidative metabolism. Of significance is the fact that nearly one quarter of the residues in collagen, proline + hydroxyproline, are derived from α-ketoglutarate. Alpha-ketoglutarate also serves as a cofactor for the hydroxylation of proline and lysine and thus its availability is crucial for collagen synthesis. It is easy to understand, then, that inhibitors of oxidative metabolism, such as the NAD analogues 6-aminonicotinamide (AN) and 3-acetylpyridine, are teratogenic[84-89] and the teratogenic organophosphorus and methylcarbamate insecticides interfere with oxidative metabolism and lower NAD levels[90]. This idea is further supported by the observation that the Kreb's cycle metabolites pyruvate, succinate and α-ketoglutarate alleviated the teratogenic effects of the above antimetabolites[91]. These studies showed that many other compounds which promote oxidative metabolism prevented or decreased the teratogenic action of NAD analogues. Ascorbate was also found to have a protective effect[91, 92]. The observation that cartilage from achondroplastic rabbits metabolises glucose at a greater rate than normal cartilage[93] is consistent with our hypothesis, as is the development of congenital anomalies in rats subjected to decreased cellular oxidative energy metabolism during the period of organogenesis, by experimental riboflavin deficiency[95, 96] and by antimycin A[97].

We have examined the metabolic regulation of collagen synthesis in primary cultures of chick chondrocytes by altering the contributions of glycolytic and Kreb's pathways[98]. These studies confirmed the relationship between oxidative metabolism and collagen synthesis. Chondrocytes in culture metabolize largely by the glycolytic pathway. Inhibition of glycolysis with iodoacetate would increase the cells' reliance on oxidative metabolism.

Iodoacetate increased the proportion of proline incorporated into collagen. Similar increases in collagen synthesis were seen when pyruvate or α-ketoglutarate concentrations were increased in the medium. The increase in collagen synthesis was manifested both in terms of increased incorporation into the collagen sequence as assayed by the use of collagenase and by the increased formation of radioactive hydroxyproline. Examination of the nature of the collagen chains synthesized in the presence of the stimulating agents showed that only the cartilage specific Type II collagen was synthesized under all conditions. Our studies confirm the role of metabolic factors in regulating collagen synthesis and suggest that the modulation of collagen synthesis may be linked to the process of teratogenesis.

Inhibition of prolyl hydroxylase

The hydroxylation of proline by prolyl hydroxylase is a critical step in collagen biosynthesis, since hydroxyproline stabilizes the collagen triple helix and higher-order aggregates by participating in hydrogen bonds[99,100], raises the denaturation temperature of collagen above the physiological (37 °C) and reduces its susceptibility to non-specific proteolysis[101,102]. Under-hydroxylated collagen is secreted inefficiently from the cells[80,103]. Inhibition of the reaction catalysed by prolyl hydroxylase can thus be expected to contribute to disturbances in morphogenesis. Interference with hydroxyproline synthesis may be an important pathway to teratogenesis, since a large variety of chemically unrelated teratogenic agents are known to inhibit prolyl hydroxylase (Table 3.3).

Table 3.3 Prolyl hydroxylase inhibitors which interfere with morphogenesis

Metals	Mercury
	Methyl mercury
	Cadmium
	Zinc
Drugs	Salicylates
	Aspirin
	Dilantin (diphenyl hydantoin)
	Hydralazine (l-hydrazinophthalazine)
	Thalidomide
	Epinephrine
Chelating agents	Bipyridyl
	EDTA

The teratogenic effects of mercury and its compounds in humans and in experimental animals are well known[104]. Mercury interacts with many biopolymers and has widespread biochemical effects. Recent studies in our laboratory have shown that prolyl hydroxylase is much more susceptible to the inhibitory effects of mercury salts than are other major biochemical processes including DNA, RNA and protein syntheses[105]. The Ki_{app} for the inhibition of prolyl hydroxylase by mercuric ions was 20 μM[106]. Since this concentration is necessary for a reduction by half in the activity of the enzyme, it may be concluded that even lower concentrations would have markedly deleterious effects on collagen synthesis.

In the recent years considerable attention has been paid to the deleterious effects of organomercurial compounds. Methyl mercury derivatives are potent teratogens in experimental animals[107-109] and in humans[110]. Methyl mercury is also a very strong inhibitor of prolyl hydroxylase with a $Ki_{app} = 15\ \mu M$[111] and its teratogenic mechanism may involve interference with collagen synthesis.

Inhibition of prolyl hydroxylase (Kiapp= 10 μM)[112] may also be a part of the teratogenic effects of cadmium compounds[113-116]. The effects of cadmium on morphogenesis are reversed by zinc[113,114]; however, zinc itself is teratogenic when administered in excess in the maternal diet of experimental hamsters[113] or rats[117]. Zinc too is an inhibitor of prolyl hydroxylase, both in tissue explants and in reactions of the purified enzyme[118].

Recent studies in our laboratory have shown that palladium is a potent inhibitor of prolyl hydroxylase (Kiapp= 20 μM), both in tissue explants and in reactions of the purified enzyme[119]. Palladium also inhibits a host of other crucial physiological processes, as seen in our *in vitro* studies[120]. In preliminary experiments in our laboratory, intra-amniotic injection of as little as 1 μg of $PdCl_2$ in rat fetuses between days 12-14 of gestation resulted in fetal resorption. Control fetuses in the opposite horn of the uterus, which were injected with normal saline, did not reveal any disturbances in morphogenesis.

Several common drugs with teratogenic potential are inhibitors of prolyl hydroxylase. Aspirin (acetylsalicylic acid) and other salicylates have been recognized as teratogens in humans[121-123] and in experimental animals[124-126]. Salicylates inhibited prolyl hydroxylase in a cell-free system, presumably by chelating iron, the prosthetic metal for the enzyme[127]. The teratogenic effects of thalidomide[128,129] may also arise from the inhibition of prolyl hydroxylase by this drug[130]. Our studies[131] have shown that Dilantin (diphenylhydantoin) is an inhibitor of prolyl hydroxylase (Kiapp= 8 mM). The inhibitory effect of Dilantin is related to its ability to chelate iron. The teratogenic effects of Dilantin are well recognised[132,133]. The teratogenicity of the neuroactive catecholamines[134,135] may also be related to the ability of epinephrine to inhibit prolyl hydroxylase. Our studies showed[136] that epinephrine competes for the superoxide free radical required in the prolyl hydroxylase reaction.

Chelating agents such as EDTA[137] and bipyridyl[138] have been shown to interfere with morphogenesis. The inhibition of prolyl hydroxylase by chelating agents has been well characterized. Our studies showed that bipyridyl inhibits the hydroxylation of collagen proline and the secretion of collagen, with no effect on the elaboration and secretion of proteoglycans[80].

We have recently shown that hydroxylation of proline in polypeptide precursors of collagen is regulated by the conformation of the polypeptide substrate[139,140]. Incorporation of the proline analogue L-azetidine carboxylic acid (LACA) in place of proline would markedly alter the conformational characteristics of the collagen precursor polypeptide and thus make it an inefficient substrate for hydroxylation. LACA has been used in studies on morphogenesis and it interferes with morphogenic events[11,138,141].

The studies discussed above suggest that interference in the hydroxylation of proline may play a significant part in the teratogenic mechanism of many drugs. Since the involvement of proteoglycans in morphogenetic events is well recognized, we also investigated the biochemical mode of action of hydralazine (l-hydrazinophthalazine) in promoting skeletal dysmorphism. Administration of hydralazine to one-day-old cockerels produced severe skeletal defects[142]. Hydralazine inhibits the synthesis of hydroxyproline in tissues, by inhibiting prolyl hydroxylase[143,144]. The inhibitory action of hydralazine on collagen synthesis and secretion was reversed by the addition of iron but not by manganese and hydralazine also failed to inhibit the synthesis of sulphated proteoglycans in embryonic cartilage explants[145]. These studies indicate that the dysmorphic effects of hydralazine arise from its ability to interfere with collagen rather than proteoglycan synthesis. Since there was no decrease in the incorporation of proline into the collagen sequence, only in its subsequent hydroxylation, these data support our hypothesis that interference with prolyl hydroxylase may contribute to teratogenesis.

Intracellular modification of lysine

Lysine in collagen is involved in more reactions than is any other residue. All post-translational modifications of lysine are related to the state of aggregation of collagen. Hydroxylation of lysine to hydroxylysine and its subsequent glycosylation regulate fibrillogenesis. The oxidative deamination of lysine and hydroxylysine occur extracellularly, and precede cross-linking reactions which occur spontaneously.

The hydroxylation of lysine is catalysed by lysyl hydroxylase in a reaction analogous to the hydroxylation of proline and is subject to interference by the same exogenous agents and conditions which affect the hydroxylation of proline. Very little information is available on pathological changes induced by the inhibition of lysyl hydroxylase, but it can be presumed that the mechanical strength and aggregation properties of collagen would be affected. The first demonstration of a hereditary molecular defect in collagen involved lysyl hydroxylase deficiency, seen in patients with Ehlers-Danlos Type VI syndrome[146]. This disease is characterized by multiple connective tissue and skeletal disorders, including kyphoscoliosis, hyperextensibility of joints and hyperelastic skin. Increased hydroxylysine synthesis may play a role in osteogenesis imperfecta and in skeletal disorders associated with rickets[147].

Although glycosylation of hydroxylysine residues in collagen appears to play an important structural role, very little is known about alterations in collagen introduced by defective glycosylation. Increased activity of collagen galactosyl and glucosyl transferases has been shown to be associated with the thickening of glomerular basement membrane in experimentally induced diabetes in animals[148,149]. Both transferases require manganese for their activity[21,22] and the biochemical basis of teratogenic effects of manganese deficiency[150] may include interference with these steps. This question has not been fully examined. Our studies showed that the concentration of manganese is highest in tissues of growing children; these studies also showed that tissues from achondroplastic subjects were not particularly deficient in manganese[151].

Inhibition of secretion of collagen

Collagen is an extracellular protein and its full potential as a mediator of differentiative events, as well as a structural material, can only be expressed under conditions where its secretion from the cell proceeds unhampered. Collagen is secreted by mechanisms similar to those for the secretion of other extracellular proteins. Collagen secretion is hampered, however, under conditions inhibitory to the hydroxylation of proline and lysine[80,145], presumably because unhydroxylated or underhydroxylated collagen is unable to form triple helices. Triple helical collagen is secreted more efficiently[21, 22].

Collagen secretion is inhibited by agents which disrupt microtubular function, colchicine and vinblastine[152, 153]. Both colchicine[154, 155] and vinblastine[156, 157] interfere with morphogenesis and induce multiple defects. Another microtubular disruptive agent, cytochalasin B, which has been shown to be teratogenic[158], was shown to affect the secretion of collagen. However, its cellular effects were widespread so that definite conclusions regarding collagen could not be reached[159].

Cleavage of procollagen extension peptides

The initially synthesized collagen chains are considerably larger than the chains in the functional molecule. The extension regions of procollagen have sequences quite different from the characteristic helical portion of the molecule. The individual chains are synthesized separately and the extension regions apparently facilitate the alignment of the three chains and in stabilizing them together by disulphide bridges before all the intracellular steps in post-translational processing are completed. The alignment in register of the three chains facilitates generation of the triple helix and its secretion. However, the presence of the extension peptides is a barrier to orderly fibrillogenesis. A defect in the processing of procollagen to collagen by scission of the extension peptides would lead to disorders in fibre formation and altered mechanical characteristics of the tissues. Such a defect is seen in genetically transmitted procollagen peptidase deficiency, in dermatasparaxic disorders in sheep[160], cattle[161] and, in the human, in Ehlers-Danlos Type VII syndrome[162]. The collagen fibres in dermatasparatic subjects have an unusual organization and ultrastructure[160].

Teratogenic effects of inhibitors of collagen cross-linking

Cross-link formation occurs in the last important series of chemical modifications of collagen. In the first step of cross-linking reactions, specific lysine and hydroxylysine residues in collagen already incorporated into fibrous aggregates, is oxidatively deaminated by lysyl oxidase, an extracellular, copper-requiring enzyme. The enzymatic modification of lysine or hydroxylysine results in the formation of a chemically reactive carbonyl group which spontaneously reacts with unaltered lysine or hydroxylysine residues, or with other lysine-derived carbonyl compounds on adjacent chains or in neighbouring molecules, to give rise to a network of intra- and intermolecular cross-links. The changes in the physicochemical and mechanical characteristics of collagen, as a consequence of cross-linking, are profound. The most pertinent changes in relation to teratogenesis are a decrease in solubility under physiological conditions and the achievement of a high tensile strength. It may be speculated that the teratogenic effects of cross-link inhibitors, such as lathyrogens, may arise from the failure of collagen to deposit in the appropriate aggregates during epithelio-mesenchymal interactions or from the premature removal of such aggregates, because of increased solubility in the physiological milieu. Because many morphogenetic events involve the development of mechanical stresses, failure to develop the necessary tensile strength may also contribute to teratogenesis.

The lathyrogens β-aminopropionitrile (BAPN) and amino-acetonitrile (AAN) are among the most widely used agents for experimental disruption of morphogenesis. BAPN is an irreversible inhibitor of lysyl oxidase *in vivo* and *in vitro*[164] and presumably AAN acts in a similar manner. Neither compound has any effect on collagen synthesis indicating that their teratogenic effects[165-168] may arise from their ability to prevent the formation of reactive carbonyl groups in collagen. Lysyl oxidase is also inhibited by mercury compounds[105].

Another teratogenic agent, D-penicillamine[169], interferes with collagen cross-linking by chelating copper and inhibiting lysyl oxidase[170] as well as by binding to the reactive carbonyls[171] and disrupting the aldimine cross-links before their reduction and stabilization[172]. Similar mechanisms may be involved in the teratogenic action of semicarbazide and thiosemicarbazide[173], agents which react with carbonyl groups.

The teratogenic consequences of copper deficiency[174] may also arise from interference with cross-linking of collagen, among other disrupted biochemical pathways.

DISCUSSION AND CONCLUSION

In order for differentiation and development to proceed in an orderly manner, a multitude of biochemical processes must proceed optimally and in concert. All cellular processes are subject to regulatory controls which are often highly vulnerable to exogenous influences and it is a marvel that defects in development are not more commonplace. A possible explanation for the infrequency with which teratogenesis is observed, is that defective operation of vital cellular processes may be lethal to the embryo and end in resorption. Teratogenic agents, acting at susceptible stages in development, interfere with biochemical processes which do not affect cell survival and interfere only with processes which regulate differentiation and development, and so allow anomalies to be observable during late stages of development and postnatally. Collagen and its synthesis are not essential for the survival of most cells and of many tissues, as is clear from the long term maintenance of various cell lines and organ cultures under conditions where their ability to synthesize collagen is limited by culture conditions or by the addition of collagen inhibitors. Many of these same cells produce large amounts of collagen in the tissues of their origin. Similar conclusions may be drawn from studies where collagen inhibitors are administered to experimental animals.

As discussed in our chapter, steps in collagen synthesis are particularly vulnerable to perturbing influences. Since cells may continue to proliferate under conditions where collagen synthesis is altered or where one or more post-translational modification steps are inhibited, the ensuing interference with differentiative events, imbalances in the production of different tissue components and altered interaction properties of the collagenous component of the extracellular matrix would lead to severe developmental disorders.

Interference with steps in collagen synthesis therefore constitutes a major biochemical mechanism of teratogenesis.

ACKNOWLEDGEMENTS

Author's studies discussed in this chapter were supported by various grants from the National Institutes of Health, U.S. Public Health Service and by a contract from the U.S. Environmental Protection Agency. We wish to thank Dr M. Z. Hussain for valuable discussions regarding this work. We also wish to thank Dr T. Z. Liu, Ms Kimie Kagawa, Ms Jamie McManus Long, Ms Maximita Tolentino, Ms Belma Enriquez, Mr Fredrick M. von Dohlen and Mr K. R. Sorensen for their contributions to our studies, and Mrs Valerie Willwerth for her assistance in the preparation of this manuscript.

References

1. Grobstein, C. (1961). Cell contact in relation to embryonic induction. *Exp. Cell Res. Suppl.*, **8**, 234
2. Grobstein, C. (1967). Mechanisms of organogenetic tissue interaction. *Nat. Cancer Inst. Monogr.*, **26**, 279
3. Grobstein, C. (1974). Developmental role of intercellular matrix: Retrospective and prospective. In: H. C. Slavkin and R. C. Greulich (eds.). *Extracellular Matrix Influences on Gene Expression*, pp. 9–16. (New York: Academic Press)
4. Kallman, F. and Grobstein, C. (1964). Fine structure of differentiating mouse pancreatic exocrine cells in transfilter culture. *J. Cell. Biol.*, **20**, 399
5. Kallman, F. and Grobstein, C. (1965). Source of collagen at epitheliomesenchymal interfaces during inductive interactions. *Dev. Biol.*, **11**, 169
6. Grobstein, C. and Cohen, J. (1965). Collagenase: Effect on the morphogenesis of embryonic salivary epithelium *in vitro*. *Science*, **150**, 626
7. Bernfield, M. R. (1970). Collagen synthesis during epitheliomesenchymal interactions. *Dev. Biol.*, **22**, 213
8. Bernfield, M. R. and Wessells, N. K. (1970). Intra- and extracellular control of epithelial morphogenesis. *Dev. Biol. Suppl.*, **4**, 195
9. Cohen, A. M. and Hay, E. D. (1971). Secretion of collagen by embryonic neuroepithelium at the time of spinal cord–somite interaction. *Dev. Biol.*, **26**, 578
10. Hay, E. D. and Dodson, J. W. (1973). Secretion of collagen by corneal epithelium. I. Morphology of the collagenous products produced by isolated epithelia grown on frozen-killed lens. *J. Cell Biol.*, **57**, 190
11. Coulombre, A. J. and Coulombre, J. L. (1972). Corneal development. IV. Interruption of collagen excretion into the primary stroma of the cornea with L-azetidine-2-carboxylic acid. *Dev. Biol.*, **28**, 183
12. Coulombre, J. L. and Coulombre, A. J. (1975). Corneal development. IV. Treatment of five-day-old embryos of domestic fowl with 6-diazo-5-oxo-L-norleucine (DON). *Dev. Biol.*, **45**, 291
13. Hay, E. D. (1973). Origin and role of collagen in the embryo. *Am. Zool.*, **13**, 1085
14. Trelstad, R. L. and Coulombre, A. J. (1971). Morphogenesis of the collagenous stroma in the chick cornea. *J. Cell Biol.*, **50**, 840
15. Reddi, A. H. (1974). Collagenous bone matrix and gene expression in fibroblasts. In: H. C. Slavkin and R. C. Greulich (eds.). *Extracellular Matrix Influences on Gene Expression*, pp. 619–625. (New York: Academic Press)
16. Reddi, A. H. (1976). Collagen and cell differentiation. In: G. N. Ramachandran and A. H. Reddi (eds.), pp. 449–478. (New York: Plenum Press)
17. Bornstein, P. and Ash, J. F. (1971). Cell-surface associated structural proteins in connective tissue cells. *Proc. Natl. Acad. Sci. (USA)*, **74**, 2480
18. Kosher, R. A. and Church, R. L. (1975). Stimulation of *in vitro* somite chondrogenesis by procollagen and collagen. *Nature*, **258**, 327

19. Dodson, J. W. and Hay, E. D. (1974). Secretion of collagen by corneal epithelium. Effect of the underlying substratum on secretion and polymerization of epithelial products. *J. Exp. Zool.*, **189**, 51
20. Meier, S. and Hay, E. D. (1974). Control of corneal differentiation by extracellular materials. Collagen as promoter and stabilizer of epithelial stroma production. *Dev. Biol.*, **38**, 249
21. Prockop, D. J., Berg, R. A., Kivirikko, K. I. and Uitto, J. (1976). Intracellular steps in the biosynthesis of collagen. In: G. N. Ramachandran and A. H. Reddi (eds.), *Biochemistry of Collagen*. pp. 163–273. (New York: Plenum Press)
22. Fessler, J. H. and Fessler, L. I. (1978). Biosynthesis of procollagen. *Annu. Rev. Biochem.*, **47**, 129
23. Stockdale, F., Okazaki, K., Nameroff, M. and Holtzer, H. (1964). 5-Bromodeoxyuridine: Effect on myogenesis *in vitro*. *Science*, **146**, 533
24. Abbott, J. and Holtzer, H. (1968). The loss of phenotypictraits by differentiated cells. V. The effect of 5-bromodeoxyuridine on cloned chondrocytes. *Proc. Natl. Acad. Sci. (USA)*, **59**, 1144
25. Tencer, R. and Brachet, J. (1973). Studies on the effects of bromodeoxyuridine (BUdR) on differentiation. *Differentiation*, **1**, 51
26. Levitt, D. and Dorfman, A. (1972). The irreversible inhibition of differentiation of limb-bud mesenchyme by bromodeoxyuridine. *Proc. Natl. Acad. Sci. (USA)*, **69**, 1253
27. Chaube, S. and Murphy, M. L. (1968). The teratogenic effects of the recent drugs active in cancer chemotherapy. In: D. H. M. Woollam (ed.). *Advances in Teratology*, 3, pp. 181–237. (New York: Academic Press)
28. Mayne, R., Vail, M. S. and Miller, E. J. (1975). Analysis of changes in collagen biosynthesis that occur when chick chondrocytes are grown in 5-bromo-2-deoxyuridine. *Proc. Natl. Acad. Sci. (USA)*, **72**, 4511
29. Daniel, J. C. (1976). Changes in type of collagen synthesized by chick fibroblasts *in vitro* in the presence of 5-bromodeoxyuridine. *Cell Differentiation*, **5**, 247
30. Hata, R.-I. and Peterkofsky, B. (1977). Specific changes in the collagen phenotype of BALB 3T3 cells as a result of transformation by sarcoma viruses or a chemical carcinogen. *Proc. Natl. Acad. Sci. (USA)*, **74**, 2933
31. Rabson, A. S., Stern, R., Tralka, T. S., Costa, J. and Wilezek, J. (1977). Hexamethylene bisacetamide induces morphological changes and increased synthesis of procollagen in cell line from glioblastoma multiforme. *Proc. Natl. Acad. Sci. (USA)*, **74**, 5060
32. Little, C. D., Church, R. L., Miller, R. A. and Ruddle, F. H. (1977). Procollagen and collagen produced by a teratocarcinoma-derived cell line TSD4: Evidence for a new form of collagen. *Cell*, **10**, 287
33. Smith, D. B., Martin, G. R., Miller, E. J., Dorfman, A. and Swarm, R. (1975). Nature of the collagen synthesized by a transplanted chondrosarcoma. *Arch. Biochem. Biophys.*, **166**, 181
34. Orkin, R. W., Gehron, P., McGoodwin, E. B., Martin, G. R., Valentine, T. and Swarm, R. (1977). A murine tumor producing a matrix of basement membrane. *J. Exp. Med.*, **145**, 204
35. Moro, L. and Smith, B. D. (1977). Identification of collagen $\alpha 1$ (I) trimer and normal Type I collagen in a Polyoma virus-induced mouse tumor. *Arch. Biochem. Biophys.*, **182**, 33
36. London, W. T., Levitt, N. H., Kent, S. G. and Sever, J. L. (1973). The teratologic effects of Venezuelan equine encephalitis virus in Rhesus monkeys. *Teratology*, **7**, A-22 (Abstract)
37. Kibrick, S. and Benirschke, K. (1958). Severe generalized disease encephalohepatomyocarditis occurring in the new born period and due to infection with Coxsackie Virus, group B: Evidence of intrauterine infection with this agent. *Pediatrics*, **22**, 857
38. Evans, T. W. and Brown, G. C. (1963). Congenital anomalies and virus infections. *Am. J. Obstet. Gynecol.*, **87**, 749
39. Henshaw, J. B. (1970). Developmental abnormalities associated with congenital Cytomegalus infection. In: Woollam, D. H. M. (ed.). *Advances in Teratology*, 4, pp. 62–93. (New York: Academic Press)
40. Brown, G. C. and Karunas, R. S. (1972). Relationship of congenital anomalies and maternal infection with selected enteroviruses. *Am. J. Epidemiol.*, **95**, 207
41. Heath, H. D., Shear, H. H., Imagawa, D. T., Jones, M. H. and Adam, J. M. (1956). Tera-

togenic effects of Herpes simplex, vaccinia, influenza-A (NWS) and distemper virus infections on early chick embryos. *Proc. Soc. Exp. Biol. Med.*, **92**, 675
42. Brown, G. C. (1966). Recent advances in the viral aetiology of congenital anomalies. In: Woollam, D. H. M. (ed.). *Advances in Teratology*, **1**, pp. 55–80. (London: Logos Press)
43. Micheals, R. H. and Mellin, G. W. (1960). Prospective experience with maternal rubella and associated congenital malformations. *Pediatrics*, **26**, 200
44. Cooper, L. Z. and Krugman, S. (1967). Clinical manifestations of postnatal and congenital rubella. *Arch. Ophthalmol.*, **77**, 434
45. Heggie, A. D. (1977). Growth inhibition of human embryonic and fetal rat bones in organ culture by Rubella virus. *Teratology*, **15**, 47
46. Bhatnagar, R. S., Hussain, M. Z., Streifel, J. A., Tolentino, M. and Enriquez, B. (1978). Alteration of collagen synthesis in lung organ cultures by hyperoxic environments. *Biochem. Biophys. Res. Commun.*, **83**, 392
47. Hussain, M. Z., Tolentino, M. and Bhatnagar, R. S. (1979). Increased Type III collagen synthesis in lung organ cultures exposed to paraquat. (Submitted for publication)
48. Ferm, V. H. (1964). Teratogenic effects of hyperbaric oxygen. *Proc. Soc. Exp. Biol. Med.*, **116**, 975
49. Khera, K. S., Whitta, L. C. and Clegg, D. J. (1970). Embryopathic effects of diquat and paraquat. In: W. B. Deichmann, J. L. Radoinski and R. A. Penalver (eds.). *Pesticides Symposia, Inter American Congress on Toxicology and Occupational Medicine*, pp. 257–261. (Miami: Halds and Associates, Inc.)
50. Fridovich, I. (1977). Biological aspects of superoxide radical and superoxide dismutases. In: O. Hayaishi and K. Asada (eds.). *Biochemical and Medical Aspects of Active Oxygen*, pp. 171–181. (Baltimore: University Park Press)
51. Ledwith, A. (1977). Electron transfer reactions of paraquat. In: A. P. Autor (ed.). *Biochemical Mechanisms of Paraquat Toxicity*, pp. 21–38. (New York: Academic Press)
52. Bhatnagar, R. S. (1977). The role of superoxide in oxidant-induced pulmonary fibrosis. In: S. D. Lee (ed.). *Biochemical Effects of Environmental Pollutants*, pp. 47–58. (Ann Arbor: Ann Arbor Science Publishers)
53. Fraser, F. C. (1969). Gene–environment interactions in the production of cleft palate. In: H. Nishimura and J. R. Miller (eds.). *Methods for Teratological Studies in Animals and Man*, pp. 34–49. (Tokyo: Igaku Shoin Ltd.)
54. Dostal, M. (1971). Morphogenesis of cleft palate induced by exogenous factors. III. Intra-amniotic application of hydrocortisone in mice. *Teratology*, **4**, 63
55. Chaudhary, A. P. and Shah, R. M. (1973). Estimation of hydrocortisone dose and optimal gestation period for cleft palate induction in golden hamsters. *Teratology*, **8**, 139
56. Bongiovanni, A. M. and McPadden, A. J. (1960). Steroids during pregnancy and possible fetal consequences. *Fertil. Steril.*, **11**, 181
57. Nanda, R. (1970). The role of sulfated mucopolysaccharides in cleft palate production. *Teratology*, **3**, 237
58. Walker, B. E. (1971). Induction of cleft palate in rats with anti-inflammatory drugs. *Teratology*, **4**, 39
59. Kivirikko, K. I., Laitinen, O., Aer, J. and Halme, J. (1965). Studies with ^{14}C-proline on the action of cortisone on the metabolism of collagen in the rat. *Biochem. Pharmacol.*, **14**, 1445
60. Blumenkrantz, N. and Asboe-Hansen, G. (1976). Cortisol effect on collagen biosynthesis in embryonic explants and *in vitro* hydroxylation of protocollagen. *Acta Endocrinol.*, **83**, 665
61. Houck, J. C., Patel, Y. M. and Gladner, J. (1967). The effects of anti-inflammatory drugs upon the chemistry and enzymology of rat skin. *Biochem. Pharmacol.*, **16**, 1099
62. Saarni, H. (1977). The effect of certain anti-inflammatory steriods on collagen synthesis *in vitro*. *Biochem. Pharmacol.*, **26**, 1961
63. Temin, H. (1965). The mechanism of carcinogenesis by avian sarcoma viruses. I. Cell multiplication and differentiation. *J. Natl. Cancer Inst.*, **35**, 679
64. Green, H., Todaro, G. J. and Goldberg, B. (1966). Collagen synthesis in fibroblasts transformed by oncogenic viruses. *Nature*, **209**, 916
65. Levinson, W., Bhatnagar, R. S. and Liu, T.-Z. (1975). Loss of ability to synthesize collagen in fibroblasts transformed by Rous sarcoma virus. *J. Natl. Cancer Inst.*, **55**, 807

66. Adams, S. L., Sobel, M. E., Howard, B. H., Olden, K., Yamada, K. M., deCrumbrugghe, B. and Pastan, I. (1977). Levels of translatable mRNAs for cell surface protein, collagen precursors and two membrane proteins are altered in Rous sarcoma virus-transformed chick embryo fibroblasts. *Proc. Natl. Acad. Sci. (USA)*, **74**, 3399
67. Dagg, C. P., Karnofsky, D. A., Lacon, C. and Roddy, J. (1956). Comparative effects of 6-diazo-5-oxo-L-norleucine and O-diazo acetyl-L-serine on the chick embryo. *Proc. Am. Assoc. Cancer Res.*, **2**, 101
68. Murphy, M. L., Dagg, C. P. and Karnofsky, D. A. (1957). Comparison of teratogenic chemicals in the rat and chick embryos. *Pediatrics*, **19**, 701
69. Thiersch, J. B. (1957). Effect of 6-diazo-5-oxo-L-norleucine (DON) on the rat litter *in utero*. *Proc. Soc. Exp. Biol. Med.*, **94**, 33
70. Aydelotte, M. B. and Kochhar, D. M. (1972). The effects of 6-diazo-5-oxo-L-norleucine (DON) on the development of mouse limb buds *in vitro*. *Teratology*, **5**, 249
71. Pratt, R. M., Goggins, J. F., Wilk, A. L. and King, C. T. G. (1973). Acid mucopolysaccharide synthesis in the secondary palate of the developing rat at the time of rotation and fusion. *Dev. Biol.*, **32**, 230
72. Greene, R. M. and Kochhar, D. M. (1975). Limb development in mouse embryos: protection against teratogenic effects of 6-diazo-5-oxo-L-norleucine (DON) *in vivo* and *in vitro*. *J. Embryol. Exp. Morphol.*, **33**, 355
73. Kochhar, D. M. (1975). The use of *in vitro* procedures in teratology. *Teratology*, **11**, 273
74. Bhatnagar, R. S. and Rapaka, S. S. R. (1971). Cellular regulation of collagen synthesis. *Nature New Biol.*, **234**, 97
75. Telser, A., Robinson, H. C. and Dorfman, A. (1965). The biosynthesis of chondroitin sulphate protein complex. *Proc. Natl. Acad. Sci. (USA)*, **54**, 912
76. Greene, R. M. and Pratt, R. M. (1977). Inhibition by diazo-oxo-norleucine (DON) of rat palatal glycoprotein synthesis and epithelial cell adhesion *in vitro*. *Exp. Cell Res.*, **105**, 27
77. Pitillo, R. F. and Hunt, D. E. (1967). In: D. Gottlieb and P. P. Shaw (eds.). *Antibiotics*, **5**, *Mechanism of Action*, pp. 481–493. (New York: Springer Verlag)
78. Ghosh, S., Blumenthal, H. J., Davidson, E. and Roseman, S. (1960). Glucosamine metabolism. V. Enzymatic synthesis of glucosamine-6-phosphate. *J. Biol. Chem.*, **235**, 1265
79. Kochhar, D. M. and Larsson, K. S. (1977). Alterations in the metabolism of glycosaminoglycans and collagen. In: J. G. Wilson and F. C. Fraser (eds.). *Handbook of Teratology*, **2**, pp. 231–269. (New York: Plenum Press)
80. Bhatnagar, R. S. and Prockop, D. J. (1966). Dissociation of the synthesis of sulfated mucopolysaccharides from the synthesis of collagen. *Biochim. Biophys. Acta*,
81. Bhatnagar, R. S. and Rapaka, R. S. (1979). Biochemical basis of teratogenesis by DON: Effect on collagen synthesis. (Submitted for publication)
82. Lane, J. M., Brighton, C. T. and Menkowitz, B. J. (1977). Anaerobic and aerobic metabolism in articular cartilage. *J. Rheumatol.*, **4**, 334
83. Reddi, A. H. and Huggins, C. B. (1971). Lactic/malic dehydrogenase quotients during transformation of fibroblasts into cartilage and bone. *Proc. Soc. Exp. Biol. Med.*, **137**, 127
84. Matschlee, G. H. and Fagerstone, K. A. (1977). Teratogenic effects of 6-aminonicotinamide in mice. *J. Toxicol. Environ. Health*, **3**, 735
85. Long, S. Y., Larsson, K. S. and Lohmander, S. (1973). Cell proliferation in the cranial base of A/J mice with 6-AN-induced cleft palate. *Teratology*, **8**, 127
86. Caplan, A. I. (1972). The effects of the nicotinamide sensitive teratogen 3-acetylpyridine on chick limb mesodermal cells in culture: Biochemical parameters. *J. Exp. Zool.*, **180**, 351
87. Landauer, W. (1957). Niacin antagonists and chick development. *J. Exp. Zool.*, **136**, 509
88. Tanaka, S., Yamamoto, Y. and Hayashi, Y. (1967). Effects of 3-acetylpyridine on the development of leg muscles of the chick embryo. *Embryology*, **9**, 306
89. Turbow, M. M., Clark, W. H. and CiPaolo, J. A. (1971). Embryonic abnormalities in hamsters following intrauterine injections of 6-aminonicotinamide. *Teratology*, **4**, 427
90. Proctor, N. H., Moscioni, A. D. and Casida, J. E. (1976). Chicken embryo NAD levels lowered by teratogenic organophosphorus and methylcarbamate insecticides. *Biochem. Pharmacol.*, **25**, 757
91. Landauer, W. and Sopher, D. (1970). Succinate, glycerophosphate and ascorbate as sources of cellular energy and as antiteratogens. *J. Embryol. Exp. Morphol.*, **24**, 187

92. Overman, D. O., Graham, M. N. and Roy, W. A. (1976). Ascorbate inhibition of 6-aminonicotinamide teratogenesis in chicken embryos. *Teratology*, 13, 85
93. Shepard, T. H. (1971). Organ-culture studies of achondroplastic rabbit cartilage: Evidence for a metabolic defect in glucose utilization. *J. Embryol. Exp. Morphol.*, 25, 347
94. Mackler, B., Grace, R., Tippit, D. F., Lemire, R. J., Shepard, T. H. and Kelley, V. C. (1975). Studies on the development of congenital anomalies in rats. III. Effects of inhibition of mitochondrial energy systems on embryonic development. *Teratology*, 12, 291
95. Aksu, O., Mackler, B., Shepard, T. H. and Lemire, R. J. (1968). Studies of the development of congenital anomalies in embryos of riboflavin-deficient, galactoflavin fed rats. II. Role of the terminal electron transport system. *Teratology*, 1, 93
96. Shepard, T. H., Lemire, R. J., Aksu, O. and Mackler, B. (1968). Studies of the development of congenital anomalies in embryos of riboflavin-deficient, galactoflavin fed rats. I. Growth and embryologic pathology. *Teratology*, 1, 75
97. Reporter, M. C. and Ebert, J. D. (1965). A mitochondrial factor that prevents the effects of antimycin A on myogenesis. *Dev. Biol.*, 12, 154
98. Bhatnagar, R. S., McManus-Long, J., Tolentino, M. and Streifel, J., (1979). Metabolic manipulation of collagen synthesis in chondrocyte cultures. (Submitted for publication)
99. Ramachandran, G. N., Bansal, M. and Bhatnagar, R. S. (1973). A hypothesis on the role of hydroxyproline in stabilizing collagen structure. *Biochim. Biophys. Acta*, 322, 166
100. Ramachandran, G. N., Bansal, M. and Ramakrishnan, C. (1975). Hydroxyproline stabilizes both intrafibrillar structure as well as inter-protofibrillar linkages in collagen. *Curr. Sci.*, 44, 1
101. Rosenbloom, J., Harsch, M. and Jimenez, S. A. (1973). Hydroxyproline content determines the denaturation temperature of chick tendon collagen. *Arch. Biochem. Biophys.*, 158, 468
102. Berg, R. A. and Prockop, D. J. (1973). The thermal transition of a non-hydroxylated form of collagen. Evidence for a role for hydroxyproline in stabilizing the triple-helix of collagen. *Biochem. Biophys. Res. Commun.*, 52, 115
103. Rapaka, R. S., Parr, R. W., Liu, T.-Z. and Bhatnagar, R. S. (1977). Biochemical basis of skeletal defects induced by hydrolazine: Inhibition of collagen synthesis and secretion in embryonic chicken cartilage *in vitro*. *Teratology*, 15, 185
104. Luckey, T. D. and Venugopal, B. (1977). Carcinogenicity and teratogenicity. In: *Metal Toxicity in Mammals*, p. 129. (New York: Plenum Press)
105. Bhatnagar, R. S., von Dohlen, F. M., Sorensen, K. R., Hussain, M. Z. and Lee, S. D. (1978). The effect of mercuric ions on connective tissue macromolecular synthesis. *Toxicol. Lett.*, 2, 217
106. Kagawa, K. J., Rapaka, R. S., Sorensen, K. R. and Bhatnagar, R. S. (1979). Inhibition of prolyl hydroxylase by mercuric ions. (Submitted for publication)
107. Gilani, S. H. (1975). Congenital abnormalities in methyl mercury poisoning. *Environ. Res.*, 9, 128
108. Gale, T. F. and Ferm, V. H. (1971). Embryopathic effects of mercuric salts. *Life Sci. Pt. 2*, 10, 1341
109. Khera, K. S. (1973). Teratogenic effects of methyl mercury on the cat: Note on the use of this species as a model for teratogenicity studies. *Teratology*, 8, 293
110. Nishimura, H. and Tanimura, T. (1976). Risk of environmental chemicals to human embryos: Mercury. In: *Clinical Aspects of Teratogenicity of Drugs*, pp. 276–280. (Amsterdam: Excerpta Medica)
111. Sorensen, K. R. and Bhatnagar, R. S. (1979). Inhibition of prolyl hydroxylase by methyl mercury. (Submitted for publication)
112. Hussain, M. Z. and Bhatnagar, R. S. (1977). Biochemical mechanisms of interaction of environmental metal contaminants with lung connective tissue. In: S. D. Lee (ed.). *Biochemical Effects of Environmental Pollutants*, pp. 341–350. (Ann Arbor: Ann Arbor Science)
113. Ferm, V. H. and Carpenter, S. J. (1967). Teratogenic effect of cadmium and its inhibition by zinc. *Nature (Lond.)*, 216, 1123
114. Ferm, V. H. (1968). The relationship between cadmium and zinc in experimental mammalian teratogenesis. *Lab. Invest.*, 18, 429

115. Chernoff, N. (1973). Teratogenic effects of cadmium in rats. *Teratology*, **8**, 29
116. Rapaka, R. S., Vare, A. M. and Bhatnagar, R. S. (1974). Biochemical mechanisms of cleft palate induction by cadmium. *J. Dent. Res.*, **53B**, 91
117. O'Dell, B. (1969). Effect of dietary components upon zinc availability: A review with original data. *Am. J. Clin. Nutr.*, **22**, 1315
118. von Dohlen, F. M., Sorensen, K. R. and Bhatnagar, R. S. (1978). Unpublished data. Submitted as part of Project Report to the U.S. Environmental Protection Agency
119. Rapaka, R. S., Sorensen, K. R., Lee, S. D. and Bhatnagar, R. S. (1976). Inhibition of hydroxyproline synthesis by palladium ions. *Biochim. Biophys. Acta*, **429**, 63
120. Liu, T. Z., Lee, S. D. and Bhatnagar, R. S. (1979). Toxicity of palladium. (Submitted for publication)
121. Richards, I. D. G. (1969). Congenital malformations and environmental influences in pregnancy. *Br. J. Prevent. Soc. Med.*, **23**, 218
122. Nelson, M. M. and Forfar, J. O. (1971). Associations between drugs administered during pregnancy and congenital abnormalities of the fetus. *Br. Med. J.*, **1**, 523
123. McNiel, J. R. (1973). The possible teratogenic effect of salicylates on the developing fetus. Brief summaries of eight suggestive cases. *Clin. Pediatr.*, **12**, 347
124. Larrson, K. S., Boström, H. and Ericson, B. (1963). Salicylate induced malformations in mouse embryos. *Acta Pediatr.*, **52**, 36
125. Warkany, J. and Takacs, E. (1959). Experimental production of congenital malformations in rats by salicylate poisoning. *Am. J. Pathol.*, **35**, 315
126. Kimmel, C. A., Wilson, J. G. and Schumacher, H. J. (1971). Studies on metabolism and identification of the causative agent in aspirin teratogenesis in rats. *Teratology*, **4**, 15
127. Nakagawa, H. and Bentley, J. P. (1971). Salicylate-induced inhibition of collagen and mucopolysaccharide biosynthesis by a chick embryo cell-free system. *J. Pharm. Pharmacol.*, **23**, 399
128. Åke Jönsson, N. (1972). Chemical structure and teratogenic properties. III. A review of available data on structure–activity relationships and mechanism of action of thalidomide analogues. *Acta Pharm. Suecica*, **9**, 521
129. Åke Jönsson, N. (1972). Chemical structure and teratogenic properties. IV. An outline of a chemical hypothesis for the teratogenic action of thalidomide. *Acta Pharm. Suecica*, **9**, 543
130. Neubert, D. (1970). Protocollagen hydroxylase in mammalian embryos and the influence of thalidomide and some of its metabolites. In: R. Bass, F. Beck, H. J. Merker, D. Neubert and B. Randhan (eds.). *Metabolic Pathways in Mammalian Embryos and then Modification by Drugs*, pp. 505–512. (Berlin: Free University Press)
131. Liu, T.-Z. and Bhatnagar, R. S. (1973). Inhibition of protocollagen proline hydroxylase by dilantin. *Proc. Soc. Exp. Biol. Med.*, **142**, 253
132. Speidel, B. D. and Meadows, S. R. (1972). Maternal epilepsy and abnormalities of the fetus and newborn. *Lancet*, **ii**, 839
133. Monson, R. R., Rosenberg, L., Hartz, S. C., Shapiro, S., Heinonen, O. P. and Slone, D. (1973). Diphenyl hydantoin and selected malformations. *N. Engl. J. Med.*, **289**, 1049
134. Jost, A., Roffi, J. and Cowitat, M. (1969). Congenital amputations determined by the BR gene and those induced by adrenalin injection in the rabbit fetus. In: C. A. Swinyard (ed.). *Limb Development and Deformity: Problems of Evaluation and Rehabilitation*, pp. 187–199. (Springfield: C. C. Thomas)
135. Goldman, A. S. and Yakovac, W. C. (1965). Teratogenic action in rats of reserpine alone and in combination with salicylate and immobilization. *Proc. Soc. Exp. Biol. Med.*, **118**, 857
136. Liu, T.-Z. and Bhatnagar, R. S. (1973). Mechanism of hydroxylation of proline. *Fed. Proc.*, **32**, 613
137. Swenerton, H. and Hurley, L. S. (1971). Teratogenic effects of a chelating agent and their prevention by Zn. *Science*, **173**, 62
138. Hilfer, S. R. and Pakstis, G. L. (1977). Interference with thyroid histogenesis by inhibitors of collagen synthesis. *J. Cell. Biol.*, **75**, 446
139. Rapaka, R. S., Renugopalakrishnan, V., Urry, D. W. and Bhatnagar, R. S. (1978). Hydroxylation of proline in polytripeptide models of collagen: Stereochemistry of polytripeptide-prolyl hydroxylase interaction. *Biochemistry*, **17**, 2892

140. Bhatnagar, R. S., Renugopalakrishnan, V., Rapaka, R. S. and Urry, D. W. (1979). Interaction between prolyl hydroxylase and its polypeptide substrate. In: *Proceedings of the International Symposium on Biomolecular Structure, Conformation, Function and Evolution, Madras, India.* (In press)
141. Aydellotte, M. B. and Kochhar, D. M. (1972). Development of mouse limb buds in organ culture: chondrogenesis in the presence of a proline analog, L-azetidine-2-carboxylic acid. *Dev. Biol.,* 28, 191
142. Comens, P. (1960). Chronic intoxication from hydralazine resembling disseminating lupus erythematosus and its apparent reversal by manganese. In: M. J. Seven (ed.). *Metal Binding in Medicine,* pp. 312–320. (Philadelphia: Lippincott)
143. Bhatnagar, R. S., Rapaka, S. S. R., Liu, T.-Z. and Wolfe, S. M. (1972). Hydralazine-induced disturbances in collagen biosynthesis. *Biochim. Biophys. Acta,* 271, 125
144. Chen, K. H., Paz, M. A. and Gallop, P. M. (1977). Collagen prolyl hydroxylation in W1-38 fibroblasts cultures: Action of hydralazine. *In Vitro,* 13, 49
145. Rapaka, R. S., Parr, R. W., Liu, T.-Z. and Bhatnagar, R. S. (1977). Biochemical basis of skeletal defects induced by hydralazine: Inhibition of collagen synthesis and secretion in embryonic chicken cartilage *in vitro. Teratology,* 15, 185
146. Pinnell, S. R., Krane, S. M., Kenzora, J. E. and Glimcher, M. J. (1972). A heritable disorder of connective tissue: Hydroxylysine deficient collagen disease. *N. Engl. J. Med.,* 286, 1013
147. Eastoe, J. E., Martens, P. and Thomas, N. R. (1973). The amino acid composition of human hard tissue collagens in osteogenesis imperfecta and dentinogenesis imperfecta. *Calcif. Tissue Res.,* 3, 49
148. Haag, E. (1974). Glomerular basement membrane thickening in rats with long-term alloxan diabetes. A quantitative electron microscope study. *Acta Pathol. Microbiol. Scand.,* 82, 211
149. Spiro, R. G. and Spiro, M. J. (1971). Studies on the biosynthesis of the hydroxylysine-linked disaccharide unit of basement membrane and collagens. III. Tissue and subcellular distribution of glycosyl-transferases and the effect of various conditions on the enzyme levels. *J. Biol. Chem.,* 246, 4919
150. Asling, C. W. and Hurley, L. S. (1963). The influence of trace elements on the skeleton. *Clin. Orthop.,* 27, 213
151. Schor, R. A., Prussin, S. G., Jewett, D. L., Ludowieg, J. J. and Bhatnagar, R. S. (1973). Trace levels of manganese, copper and zinc in rib cartilage as related to age in humans and animals, both normal and dwarfed. *Clin. Orthop.,* 93, 346
152. Dehm, P. and Prockop, D. J. (1972). Time lag in the secretion of collagen from matrix-free tendon cells and inhibition of the secretory process by colchicine and vinblastine. *Biochim. Biophys. Acta,* 264, 375
153. Diegelmann, R. F. and Peterkofsky, B. (1972). Inhibition of collagen secretion from bone and cultured fibroblasts by microtubular disruptive drugs. *Proc. Natl. Acad. Sci. (USA),* 69, 892
154. Ferm, V. H. (1963). Colchicine teratogenesis in hamster embryos. *Proc. Soc. Exp. Biol. Med.,* 112, 775
155. Didcock, K., Jackson, D. and Robson, J. M. (1956). The action of some nucleotoxic substances on pregnancy. *Br. J. Pharmacol.,* 11, 437
156. Ferm, V. H. (1963). Congenital malformations in hamster embryos after treatment with vinblastine and vincristine. *Science,* 141, 426
157. Demeyer, W. (1964). Vinblastine-induced malformations of face and nervous system in two rat strains. *Neurology,* 14, 806
158. Linville, G. P. and Shepard, T. H. (1972). Neural tube closure defects due to cytochalasin B. *Nature,* 236, 246
159. Bornstein, P. (1974). The biosynthesis of collagen. *Annu. Rev. Biochem.,* 43, 567
160. Fjølstad, M. and Helle, O. (1974). A hereditary dysplasia of collagen tissues. *J. Pathol.,* 112, 183
161. O'Hara, P. J., Read, W. K., Romane, W. M. and Bridges, C. H. (1970). A collagenous tissue dysplasia of calves. *Lab. Invest.,* 23, 307
162. McKusick, V. A. (1972). The Ehlers-Danlos Syndrome. In: *Heritable Disorders of Connective Tissue,* 4th Ed. p. 292. (St. Louis: C. V. Mosby)

163. Duksin, D. and Bornstein, P. (1977). Impaired conversion of procollagen to collagen by fibroblasts and bone treated with tunicamycin, an inhibitor of protein glycosylation. *J. Biol. Chem.*, **252**, 955
164. Siegel, R. C., Pinnell, S. R. and Martin, G. R. (1970). Cross-linking of collagen and elastin. Properties of lysyl oxidase. *Biochemistry*, **9**, 4486
165. Abramovich, A. and Devoto, F. C. H. (1968). Anomalous maxillofacial patterns produced by maternal lathyrism in rat foetuses. *Arch. Oral Biol.*, **13**, 823
166. Hall, B. K. (1972). Skeletal defects in embryonic chicks induced by administration of beta-aminopropionitrile. *Teratology*, **5**, 81
167. Pratt, R. M. and King, C. T. G. (1972). Inhibition of collagen cross-linking associated with β-aminopropionitrile-induced cleft palate in the rat. *Dev. Biol.*, **27**, 322
168. Wiley, M. J. and Joneja, M. G. (1976). The teratogenic effects of β-aminopropionitrile in hamsters. *Teratology*, **14**, 43
169. Merker, H. J., Franke, L. and Günther, T. (1975). The effect of D-penicillamine on the skeletal development of rat foetuses. *Naunyn-Schmiedberg's Arch. Pharmacol.*, **287**, 359
170. Nimni, M., Deshmukh, K. and Gerth, N. (1972). Collagen defect induced by penicillamine. *Nature*, **240**, 220
171. Deshmukh, K. and Nimni, M. (1969). A defect in the intramolecular and inter-molecular cross-linking of collagen caused by penicillamine. *J. Biol. Chem.*, **244**, 1787
172. Nimni, M. E., Deshmukh, K. and Deshmukh, A. (1970). Mechanism of inhibition of collagen cross-linking and depolymerization of an incompletely cross-linked form of insoluble collagen caused by D-penicillamine. In: E. Balazs (ed.). *Chemistry and Molecular Biology of the Intercellular Matrix*, 1, pp. 417–431. (New York: Academic Press)
173. Neuman, R. E., Maxwell, M. and McCoy, T. A. (1956). Production of beak and skeletal malformations of chick embryos by semicarbazide. *Proc. Soc. Exp. Biol. Med.*, **92**, 578
174. Everson, G. J., Shrader, R. E. and Wang, T. (1968). Chemical and morphological changes in brains of copper-deficient guinea pigs. *J. Nutr.*, **96**, 115

4
Biochemical and ultrastructural aspects of [^{14}C]glucosamine utilization by normal and alkaloid treated preimplantation mouse blastocysts

DANICA DABICH, R. A. ACEY AND LINDA D. HAZLETT

INTRODUCTION

One of the enigmas in mammalian developmental biology has been the reconciliation of the comparatively long duration of the preimplantation period with the lack of significant growth[1] or *apparent* developmental change during this period. The validity of the latter concept, however, is being increasingly challenged as a consequence of technological improvements which are applicable to microsystems and which can detect molecular[2] and/or ultrastructural changes in preimplantation embryos. By perturbing normal developmental events with teratogenic or embryopathic agents, by genetic mutation, or by formation of chimeras[3], one is provided with additional means of probing normal versus abnormal development and clarifying intricate developmental processes. It was with these thoughts in mind that the present studies were approached.

The preimplantation mouse blastocyst was examined because of its pivotal role in development[1]. Since the adhesive phase of implantation appears to be functionally linked to biochemical changes in trophoblast surface membranes[4-10] and since amino sugars are found in membrane glycoproteins[11] and glycolipids[12], [^{14}C]glucosamine was used to pulse label preimplantation mouse blastocysts incubated in the presence and absence of vinblastine or colchicine. These alkaloids impair microtubule function[13] and through coordinated and/or cooperative cellular effects cause surface modifications[14-16]. This report, therefore, describes attempts to evaluate normal blastocyst development and the effects of alkaloids, in particular vinblastine, on preimplantation blastocysts through: (i) [^{14}C]glucosamine uptake and incorporation measurements, (ii) vinblastine localization of ^{14}C-labelled products, and (iii) vinblastine-induced alterations in ultrastructure and surface morphology.

MATERIALS AND METHODS

Procedures for superovulation[17], *in vitro* blastocyst culture from the two cell stage[17], measurements of uptake and incorporation of [^{14}C]glucosamine[18] or ^{14}C-amino acids[19] by blastocysts, and radioautography of blastocysts[18] have been described elsewhere. Only experimental details specific to this study or not previously reported are presented here.

Early blastocysts [88–92 h post coitum (p.c.)] were obtained from inbred Swiss Webster mice between 17 and 19 h on Day 4 of pregnancy[18] whereas late blastocysts (98–102 h p.c.) were obtained between 3 and 5 h on Day 5 of pregnancy[18]. Embryos for each experimental group were random samples obtained from a common pool of blastocysts of the appropriate developmental stage. Individual samples within a single experiment were comprised of groups of 10 embryos each unless indicated otherwise. Culture medium (modified Brinster's medium[18], MBM) containing vinblastine (Eli Lilly; Indianapolis, In), colchicine (recrystallized, Sigma Chem. Co.; St. Louis, Mo) or lumicolchicine (an ultraviolet irradiation product of colchicine) was prepared by dilution of a stock solution of the appropriate drug (100 μg/ml 3% (v/v) ethanol) with MBM to produce the required final drug concentrations. Statistical analysis of data resulting from quantitative biochemical measurements was performed with the Kruska-Wallis[20] test alone or in conjunction with analysis of variance[21] or a multiple range test[21].

Initial investigations indicated that seasonal variation appears to affect the absolute numerical values measured for total uptake and incorporation of precursor. Each individual experiment was, therefore, always performed concurrently with the appropriate controls. Although absolute values for specific biochemical parameters did change seasonally, the overall trend or observation was repeatable.

Scanning electron microscopy (SEM) was performed with late blastocysts which had developed *in utero* to this stage and which were fixed after two hours of incubation *in vitro* in the presence or absence of vinblastine (0.10 ng/ml medium). Cultured embryos (control and experimental groups) were washed once in warm MBM[18] prior to preparation for SEM. Embryos from control and experimental groups, respectively, were collected in a minimum volume of MBM and transferred directly onto Millipore membrane filters (15 mm diameter) which had been suspended at the apex of 15 × 100 mm test tubes and just barely covered with cold (4 °C) fixative solution, 1% OsO_4–6.5% (v/v) glutaraldehyde in Sorensen's phosphate buffer, pH 7.4. Embryos were forced onto the membrane by gentle centrifugation, 800 × g for 7 min, 4 °C, then fixed for 40 minutes. The membranes were dehydrated in graded concentrations of ethanol (50–100%) and critical point dried in liquid carbon dioxide. The membranes were mounted on aluminium stubs, then gold-coated (200–500 Å) in a Hummer evaporator. An ETEC Autoscan scanning electron microscope at 20 KV was used for viewing the samples. Polaroid black and white positive/negative 55 film was used for photography.

RESULTS

Pulse labelling of early blastocysts *in vitro*

Results of quantitative measurements of total uptake of radioactivity and subsequent incorporation of [^{14}C]glucosamine into acid precipitable substances of normal preimplantation blastocysts are presented in Figure 4.1.

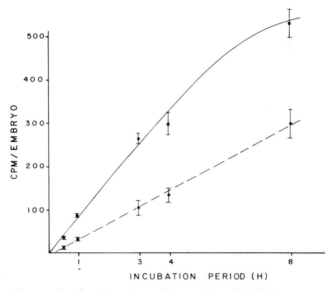

Figure 4.1 Time profile of total uptake, and incorporation of early blastocysts pulse labelled with [^{14}C]glucosamine *in vitro*. The final glucosamine concentration was 15.1 μM (specific activity 318 μCi/μmole). Each experimental point represents the average value $\pm$ S.E.M. obtained from three to five independent experiments with three to five samples per experiment. The regression line for the linear portion of the curve was obtained from an unweighted least squares programme[22]. The r values for uptake and incorporation, are 0.98 and 0.99, respectively. CPM denotes counts per minute, whereas ●———● refers to uptake, and ●---● to incorporation

Since both total uptake and incorporation of radioactivity represent the summation of several biochemically distinct steps, these values should not be equated to intracellular transport rates of the precursor[23] or absolute macromolecular synthetic rates[19], respectively. Total uptake in these studies represents the amount of isotope retained by the embryos as both acid soluble and acid precipitable substances. Incorporation refers only to the amount of radioactivity accumulated by the embryos in acid precipitable form after a finite period of pulse labelling. It is, therefore, a composite function of the specific activity of the endogenous precursor pool and the absolute synthetic versus catabolic rates[24,25] for the total spectrum of ^{14}C-labelled molecules. Detectable incorporated radioactivity is exclusively glucosamine whereas the majority of the acid soluble radioactivity is a phosphorylated derivative of glucosamine with trace amounts of the UDP-N-acetyl glucosamine[26]. Quantitative measurements of both total uptake and incorporation, moreover, include radioactive precursors and/or products present in the blastocoele. For the purposes of this study, measurements of total uptake and incorporation with normal early and late preimplantation blastocysts provide the background against which the developmental changes in blastocysts and effects of alkaloids on such embryos are assessed.

With subsaturating amounts of isotope, the time course of total uptake of the [^{14}C]glucosamine by early blastocysts was linear for the first three hours of pulse labelling whereas incorporation proceeded linearly for 8 h under identical conditions (Figure 4.1). A lag period between uptake of the isotope and its appearance in acid precipitable material is indicated by the fact that the line describing the best fit for the data[22] does not pass through the origin. This suggests that there may be a large precursor pool of glucosamine or its derivatives, and/or that the ^{14}C-labelled precursor and its derivatives do not readily equilibrate with the kinetic pool(s) of precursor used for biosynthesis. The lag in incorporation of [^{14}C]glucosamine by blastocysts is analogous to studies of ^{14}C-amino acid incorporation, a parameter believed to be affected by cellular compartmentalization of different kinetic pools[27, 28]. The analogy is

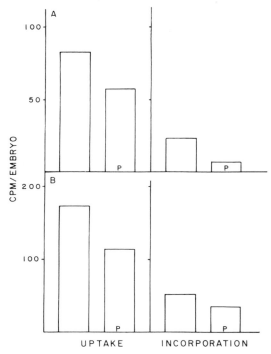

Figure 4.2 Effect on total uptake and incorporation as a result of preloading early blastocysts with non-labelled glucosamine prior to pulse labelling with [^{14}C]glucosamine. Embryos were cultured for 1 h in 1.5 mM non-radioactive glucosamine prior to pulse labelling with [^{14}C]glucosamine (final concentration = 47.6 μM; specific activity = 200 μCi/μmole). A = samples pulse labelled for 0.5 h; B = samples pulse labelled for 1 h. Results are expressed as cpm/embryo and each data point represents the average of three experiments. Each experimental determination was performed with 25–50 embryos per experiment. P denotes preincubated embryos

[^{14}C]GLUCOSAMINE UPTAKE BY NORMAL AND DRUG TREATED BLASTOCYSTS

strengthened by results of experiments involving preloading of blastocysts with saturating levels of unlabelled glucosamine prior to pulse labelling (see Figure 4.2 and Table 4.1). After one half-hour of labelling, there is a wide disparity in the total uptake/incorporation (U/I) ratio of preloaded embryos compared to controls, i.e. 9.5 versus 3.6, respectively. Of the two parameters in this ratio, incorporation is more severely affected than uptake. After one hour the ratio is like that of the controls, although the absolute values of both parameters are still depressed. The effect of preloading on the U/I ratio might, therefore, be envisioned to result from saturation of an endogenous kinetic pool with significant amounts of unlabelled glucosamine metabolite.

Table 4.1 Effect of preloading on uptake/incorporation ratios of pulse labelled early blastocysts

Incubation period (h)	Uptake/incorporation	
	Control	Preincubated
0.5	3.6	9.5
1.0	3.4	3.2

Calculation of these values is based on data in Figure 4.2

Pulse labelling of late blastocysts *in vitro*

Under identical labelling conditions used for early blastocysts, both total uptake and incorporation of [^{14}C]glucosamine are significantly elevated in late versus early blastocysts ($p < 0.001$) (Figure 4.3). Statistically significant

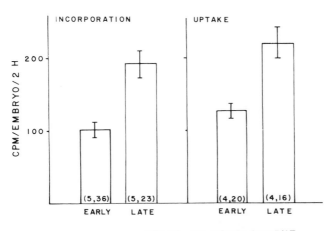

Figure 4.3 Total uptake and incorporation of [^{14}C]glucosamine by early and late blastocysts labelled *in vitro*. Statistically significant differences[20] exist between early and late stage blastocysts with respect to both total uptake ($p < 0.01$) and incorporation ($p < 0.01$). The number of individual experiments and total number of samples, respectively, is indicated in parentheses. Final [^{14}C]glucosamine concentration (specific activity = 318 μCi/μmole) was 12.9 μM. Results as [CPM/embryo/2 h] $\pm$ 95% confidence limits

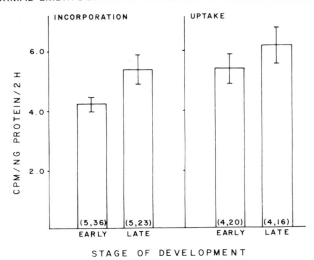

Figure 4.4 Total uptake and incorporation of [^{14}C]glucosamine based on total protein content of early and late blastocysts. Calculations are based on literature values for total proetin content of embryos; 23 ng for early blastocysts and 35 ng for late blastocysts[29]. Results are expressed as [CPM/ng protein/2 h] ± 95% confidence limits. The amount of incorporated radioactivity was significantly elevated[20] in late versus early blastocysts ($p < 0.05$) while total uptake between the two stages of development was statistically indistinguishable[20]. The numbers in parentheses indicate the number of individual experiments and samples, respectively. Final [^{14}C]glucosamine concentration (specific activity = 318 μCi/μmole) was 12.9 μM

differences[20] in incorporation between the two developmental stages are found regardless of whether this parameter is expressed as cpm/embryo or cpm/total embryonic protein (see Figures 4.3 and 4.4). On the other hand, normalization of the values for total uptake on the basis of total embryonic protein indicates that there is no statistical difference between early versus late blastocysts (Figure 4.4). Thus total uptake stays abreast of increasing protein mass of late blastocysts. The increased incorporation in late versus early blastocysts may have resulted from changes in any one, or combinations, of the following parameters: increased absolute rates of synthesis of glycosylated products; qualitative changes in the types of macromolecules expressed in late versus early blastocysts; changes in turnover rates of glycosylated products; or more rapid transport of the exogenous, labelled precursor. Of these possibilities, changes in gene expression[30,31] and precursor transport[26] have been observed between the two stages of development so far.

To focus more closely on the increased incorporation found in late versus early blastocysts, the effect of time *in utero* was examined relative to incorporation capability. Earlier studies in this laboratory by VanWinkle[24] led to the hypothesis that a change(s) occurred during blastocyst development *in utero* which caused a significant and rather abrupt increase in ^{14}C-amino acid incorporation. This hypothesis was tested relative to both ^{14}C-amino acid and [^{14}C]glucosamine incorporation. The results shown in Figure 4.5 confirm the prediction. A statistically significant ($p < 0.05$) transition[21] to a higher level of

[14C]GLUCOSAMINE UPTAKE BY NORMAL AND DRUG TREATED BLASTOCYSTS

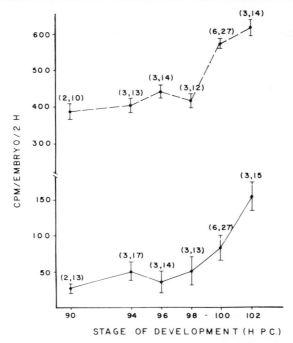

Figure 4.5 Incorporation of ^{14}C-amino acids and [^{14}C]glucosamine by blastocysts of increasing gestational age after pulse labelling *in vitro*. Results are expressed as |CPM/embryo/2 h| ± 95% confidence limits. A statistically significant increase[21] ($p < 0.05$) both in [^{14}C]glucosamine and ^{14}C-amino acid incorporation occurs between 98–100 h p.c. Final precursor concentrations were 19.1 μM (specific activity = 318 μCi/μM) for ^{14}C-glucosamine and 4.4 μCi/ml (specific activity = 58 mCi/mAtom) for the amino acid mixture. The number of individual experiments and samples, respectively, is indicated in parentheses. P.C. = post coitum, ●---● = amino acid incorporation, and ●——● = glucosamine incorporation

incorporation occurs at approximately 98 h p.c. for both the ^{14}C-labelled carbohydrate and mixed amino acids.

More importantly, the sharp increase in precursor incorporation appears to coincide with initiation of morphological changes associated with late blastocysts[32]. Therefore, changes in precursor utilization appear to correlate with observable changes in embryonic morphology. Since precursor utilization changes dramatically during a relatively short period of time, it should be possible, eventually, to pinpoint specific biochemical changes which are the result of or cause changes in blastocyst morphology or cellular physiology. With respect to incorporation of radioactive metabolites, the factors involved in the transition between early and late blastocysts appear to be under embryonic control only in part. The latter conclusion is based on the following evidence. Late embryos developed *in vitro* in Brinster's medium[17] from early embryos show normal values for both total uptake and incorporation (Figure 4.6). On the other hand, when two-cell stage embryos are cultured to late blastocysts, both total uptake and incorporation are significantly lower[20] than

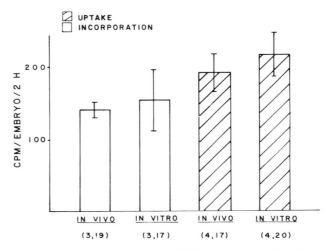

Figure 4.6 Total uptake and incorporation of [^{14}C]glucosamine by late blastocysts which developed *in vitro* from *in utero* developed early blastocysts. Results are expressed as [CPM/embryo/2 h] ± 95% confidence limits. No significant difference[20] exists between *in utero* and *in vitro* developed late blastocysts with respect to total uptake or incorporation of [^{14}C]glucosamine. Final [^{14}C]glucosamine concentration (specific activity = 200 μCi/μmole) was 19.1 μM. The number of individual experiments and samples, respectively, is indicated in parentheses

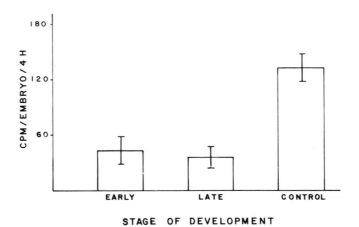

Figure 4.7 Incorporation of [^{14}C]glucosamine by early and late blastocysts developed *in vitro* from the two-cell stage and labelled *in vitro*. Blastocysts were developed *in vitro* according to Brinster[17]. Results are expressed as [CPM/embryo/4 h] ± 95% confidence limits. Final [^{14}C]glucosamine concentration (specific activity = 318 μCi/μmole) was 15.1 μM. Control embryos are normal early blastocysts pulse labelled for 4 h

[^{14}C]GLUCOSAMINE UPTAKE BY NORMAL AND DRUG TREATED BLASTOCYSTS

normal in these blastocysts (Figure 4.7). In fact, such embryos must be pulse labelled *in vitro* for 4 hours to accumulate a significant amount of radioactivity in acid precipitable material. Therefore, at some critical stage(s) of development prior to the early blastocyst, conditions in the normal uterine environment are necessary for complete early blastocyst development judged by total uptake and/or incorporation of [^{14}C]glucosamine.

Total uptake and incorporation of [^{14}C]glucosamine by early and late blastocysts pulse labelled *in vitro* in the presence of vinblastine, colchicine or lumicolchicine

Figure 4.8 illustrates the results of attempts to determine the minimum concentration of vinblastine required to induce detectable quantitative changes in incorporation of [^{14}C]glucosamine by early and late blastocysts. Incorporation of the precursor by late, but not early, blastocysts is diminished significantly[22] ($p < 0.05$) when vinblastine is present at a final concentration of either 0.1 ng/ml or 0.5 ng/ml in the incubation medium. Furthermore, embryos incubated *in vitro* for 16 h in the presence of vinblastine at concentrations as high as 0.5 ng/ml are visibly indistinguishable under the light microscope

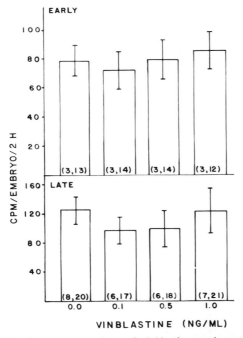

Figure 4.8 Effect of various concentrations of vinblastine on incorporation of [^{14}C]glucosamine by early and late blastocysts labelled *in vitro*. Results are expressed as [CPM/embryo/2 h] ± 95% confidence limits. Embryos were labelled with [^{14}C]glucosamine at a final concentration of 15.1 μM (specific activity = 3.18 μCi/μmole) in the presence of the appropriate vinblastine concentration. The number of individual experiments and total number of samples, respectively, is indicated in parentheses

from embryos incubated *in vitro* in the absence of the drug. Since a vinblastine concentration of 1.0 ng/ml induces gross morphological damage (initiation of blastocyst collapse), this data point was excluded from the final statistical analysis.

With late blastocysts concomitant diminution in total uptake of [^{14}C]glucosamine does not occur, at a drug concentration of 0.1 ng vinblastine/ml (Figure 4.9), nor are total uptake and incorporation of a ^{14}C-amino acid

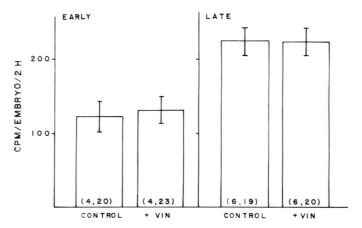

Figure 4.9 Effect of vinblastine on total uptake of [^{14}C]glucosamine by early and late blastocysts labelled *in vitro*. Results are expressed as [CPM/embryo/2 h] ± 95% confidence limits. The final vinblastine concentration was 0.1 ng/ml and [^{14}C]glucosamine 15.1 μM (specific activity = 318 μCi/μmole). The number of individual experiments and samples, respectively, is indicated in parentheses. + VIN denotes vinblastine containing samples

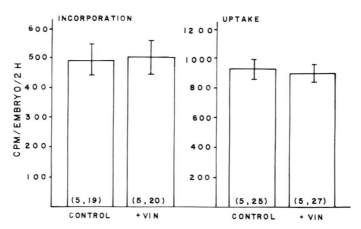

Figure 4.10 Effect of vinblastine on total uptake and incorporation of [^{14}C]amino acids. Results are expressed as [CPM/embryo/2 h] ± 95% confidence limits. The final vinblastine concentration was 0.1 ng/ml, whereas that of the amino acid mixture was 4.4 μCi/ml (specific activity = 58 mCi/mAtom). The number of individual experiments and total number of samples, respectively, is indicated in parentheses. + VIN denotes vinblastine containing samples

mixture affected by the drug (Figure 4.10). Thus, as a first approximation, it appears that neither the ability to take up precursor nor to synthesize peptide substrates for glycosylation appears to account for the diminished incorporation value. Furthermore, colchicine, which like vinblastine disrupts microtubule integrity[13], also inhibits [^{14}C]glucosamine incorporation but not total uptake of the precursor (Table 4.2). At the same concentrations,

Table 4.2 Effect of various concentrations of colchicine and lumicolchicine on total uptake and incorporation of [^{14}C]glucosamine by late blastocysts labelled *in vitro*

	Drug concentration (pg/ml)	Control	Colchicine	Lumicolchicine
		[CPM/embryo/2 h] $\pm$ 95% confidence limits		
Incorporation	50	120 $\pm$ 24 (4, 14)	124 $\pm$ 24 (4, 14) n.s.	122 $\pm$ 22 (4, 14) n.s.
	250	153 $\pm$ 23 (5, 14)	110 $\pm$ 24 (5, 16) $p < 0.05$	—
	250	118 $\pm$ 15 (4, 14)	—	124 $\pm$ 12 (4, 14) n.s.
	500	151 $\pm$ 17 (5, 19)	135 $\pm$ 22 (5, 19) n.s.	—
Uptake	250	236 $\pm$ 24 (4, 16)	232 $\pm$ 20 (4, 16) n.s.	210 $\pm$ 21 (4, 17) n.s.

Results are expressed as [CPM/Embryo/2 h] $\pm$ 95% confidence limits. Final glucosamine concentration was 24.4 μM (specific activity = 200 μCi/μmole). The number of individual experiments and the total number of samples, respectively, is indicated in parentheses. n.s. = not significant. *p* values were calculated using the Kruska-Wallis test[20]

Table 4.3 Effect of vinblastine on incorporation of [^{14}C]glucosamine by *in vitro* cultured blastocysts

Stage of development			Control	+ Vinblastine
88–92 h p.c.	98–102 h p.c.		CPM/embryo/2 h $\pm$ 95% confidence units	
Early in utero	8–10 h →		89 $\pm$ 13 (3, 13)	83 $\pm$ 13 (3, 17)
	Late in utero	8–10 h →	198 $\pm$ 22 (4, 18)	178 $\pm$ 36 (4, 18)
Early in utero	16–18 h →		158 $\pm$ 22 (4, 15)	156 $\pm$ 34 (4, 17)

The primary references are values for blastocysts which developed *in utero* (ca. 88–92 h p.c. for early blastocysts versus 98–102 h p.c. for late blastocysts). *In vitro* incubation periods are represented by horizontal arrows and the actual duration of *in vitro* incubation in the absence of vinblastine denoted numerically above the line. Incorporation was measured with late blastocysts after *in vitro* incubation in the presence of 0.1 ng vinblastine/ml incubation medium. Final [^{14}C]glucosamine concentration was 15.1 μM (specific activity = 318 μCi/μmole). The number of individual experiments and total number of samples, respectively, is cited in parentheses. p.c. = post coitum; n.s. = not significant

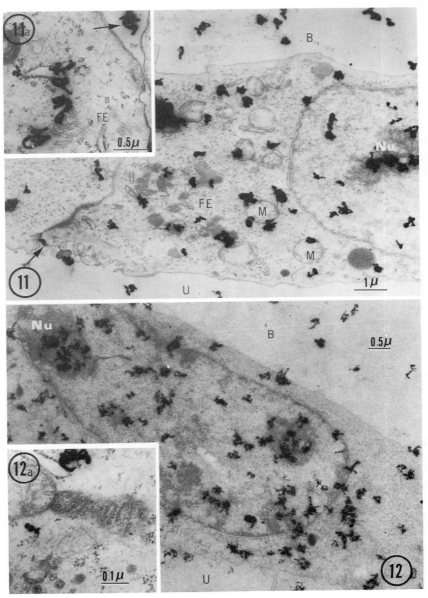

Figures 4.11–4.12A
These figures illustrate the fine structural localization of radioactivity in control and vinblastine treated trophoblast cells of late mouse blastocysts after 2 h of *in vitro* incubation in [^{14}C]-glucosamine

Figure 4.11 Nuclear as well as nucleolar (Nu) associated radioactivity is evident in a late trophoblast control. In the cytoplasm, mitochondria (M) and fibrous elements (FE) are labelled. Grains also are seen in the blastocoelic (B) and uterine (U) cavities as well as in the extracellular space between trophoblast cells (arrow). × 9600

lumicolchinine[13], a biologically inactive photo-oxidation product of colchicine used to account for non-specific drug effects, has no effect on either [^{14}C]glucosamine incorporation or uptake (Table 4.3). Thus, maintenance of microtubule integrity appears to have bearing on glycosylating activity measured by [^{14}C]glucosamine incorporation in late blastocysts.

Drug-induced perturbations of reactions leading to precursor activation, disruption in subcellular compartmentalization, qualitative changes in peptide or protein substrate synthesis and protein synthesis, and changes in turnover of cellular products (both rates and types of molecules affected) could alter values found for incorporation. That the changes brought about by the vinca alkaloid treatment are extensive and that a number of the cited factors may be involved is indicated in the ultrastructural analyses described below.

At this time the correlating studies, ^{14}C-amino acid utilization and ultrastructural examinations, have not been completed with late blastocysts treated with 0.5 and 0.1 ng vinblastine/ml. What data (biochemical and morphological) have been accumulated also indicate that the return to an apparently normal value for [^{14}C]glucosamine incorporation in the presence of 1.0 ng vinblastine/ml, is fortuitous (Figure 4.8).

In contrast to late blastocysts, early blastocysts appeared to be refractory to vinblastine treatment, as assessed by [^{14}C]glucosamine uptake and incorporation measurements (Figure 4.10). These values are indistinguishable from those of the controls. Similarly, vinblastine does not affect [^{14}C]glucosamine incorporation of late blastocysts which have either developed *in vitro* from early blastocysts or which after *in vitro* development are incubated *in vitro* for eight hours (Table 4.3). Apparently, during embryonic development from early to late blastocysts, the milieu of the uterine environment provides for developmental events which manifest themselves in the late embryo as sensitivity of glucosamine incorporation to vinblastine.

From these studies alone it is difficult to predict whether the apparent refractoriness of early blastocysts to vinblastine is due to lack of penetration of the drug, or whether binding of the drug to microtubules is affected. Further studies with radiolabelled vinblastine might facilitate solutions to these questions since ultrastructural studies with early blastocysts show no apparent subcellular or organelle disruption after drug treatment.

Ultrastructural studies: radioautography and SEM

The subcellular localization of radioactivity in normal early blastocysts after 4 h of *in vitro* incubation in the presence of [^{14}C]glucosamine was examined in

Figure 4.11A Higher magnification to better illustrate grains associated with cytoplasmic fibrous elements (FE) and in the extracellular space (arrow) in a control trophoblast. × 20 480

Figure 4.12 Vinblastine-treated trophoblast cells exhibit nuclear and nucleolar (Nu) associated label. Both the blastocoelic (B) and uterine (U) cavities also were labelled as in control preparation. However, in contrast to late controls, the trophoblast cells of vinblastine-treated embryos are characterized by a cytoplasm which appears homogeneous with few intact organelles. × 10 800

Figure 4.12A No fibrous elements are evident within the cytoplasm of any vinblastine treated blastocysts. × 48 000

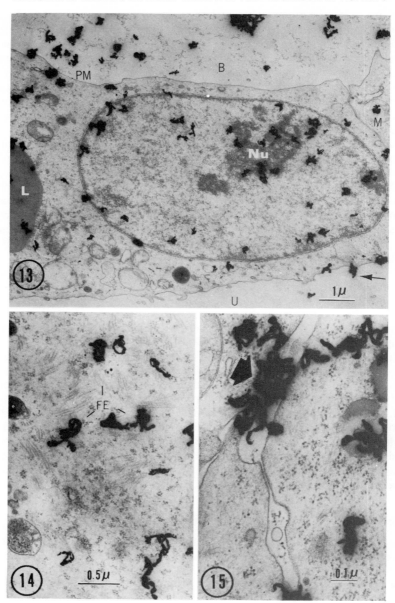

Figures 4.13–4.15
These figures illustrate the ultrastructural localization of radioactivity in vinblastine-treated trophoblast cells of early mouse blastocysts after 2 h of *in vitro* incubation in [^{14}C]glucosamine.
Control cells which have been described previously exhibit similar silver grain localizations

Figure 4.13 Silver grains are seen in association with the nucleus, nucleolus (Nu) and nuclear envelope. In the cytoplasm, label is observed over a mitochondrion (M) and a lipid droplet (L). Radioactivity is seen also within the blastocoelic (B) cavity and its abluminal plasma membrane

a series of correlative ultrastructural radioautographic studies. A slightly longer labelling period is required for clearly defined radioautograms[18]. In both the early[18] and late blastocyst (Figures 4.11, 4.11A) silver grains are most numerous within the blastocoelic cavity and appear localized on the precipitated fibrillar material which characteristically fills this lumen. Labelled components of the trophoblasts include both the blastocoelic and uterine cavity surfaces which are characterized by numerous slender microvilli. Cytoplasmic labelling is associated with lipid droplets, fibrous elements[33, 34], Golgi complexes, mitochondria, and elaborate lamellar whorls[18] which are located near the lateral border of these cells and contain an electron dense material. Numbers of grains are also observed intercellularly in areas of pinocytic activity (Figures 4.11, 4.11A).

The nuclear components which exhibit radioactive deposition include both the nucleus and nucleolus. Grains are also localized near or on the membranes of the nuclear envelope (Figures 4.11, 4.11A).

Radioautography of sections of vinblastine-treated late blastocysts (Figures 4.12, 4.12A), demonstrates localization of cellular labelling similar to normal[18] and vinblastine-treated early blastocysts (Figures 4.13–4.15) and normal late blastocysts (Figures 4.11, 4.11A). In contrast to normal late blastocysts, however, cytoplasmic labelling of drug-treated late blastocysts is decreased, whereas that of the nuclear and nucleolar regions appears to be quite dense (Figure 4.12). A striking feature is that the cytoplasm of the drug-treated cells appears 'washed out' and relatively homogeneous (Figure 4.12). No fibrous elements and their associated labelling are observed within the cytoplasm (compare Figures 4.11, 4.11A and 4.12, 4.12A). Vinblastine apparently mediated their disruption along with that of microtubules which are infrequently observed even in normal late or early embryos.

Ultrastructural autoradiographic results with vinblastine-treated early blastocysts did not exhibit discernible drug-induced changes (Figures 4.13–4.15) and therefore, supported the biochemical data indicating that these embryos may be refractory to vinblastine treatment. The results of fine structural morphological and autoradiographic experiments with drug-treated early blastocysts are virtually indistinguishable from normal control embryos[18]. Similar labelling of both the nuclear and cytoplasmic as well as surface components is found within these embryos (Figure 4.13).

Label is observed also in association with arrays of fibrous elements, which were not perturbed by drug treatment of the early blastocysts (Figure 4.14). Labelling of both the blastocoelic cavity and the surface facing the uterine cavity, as well as their respective bounding plasma membranes is seen (Figure 4.13). Additionally intercellular labelling between trophoblasts is dense (Figure 4.15).

(PM), as well as on the uterine (U) cavity surface and respective plasma membrane (arrow). × 10 240

Figure 4.14 Unlike vinblastine-treated late trophoblast cells, similarly treated early trophoblast cells show silver grains localized over fibrous elements (FE). These structures in early cells were not perturbed by drug treatment. × 22 400

Figure 4.15 As in control early trophoblast cells, numerous silver grains are observed in the extracellular space between trophoblast cells of vinblastine-treated early cells. × 36 000

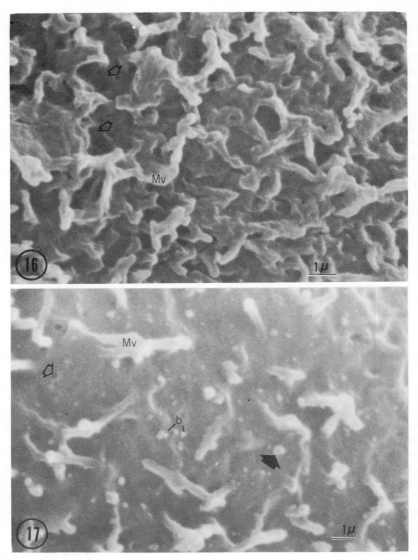

Figures 4.16 and 4.17
These figures consist of scanning electron micrographs of control and drug treated late mouse blastocysts

Figure 4.16 Note numerous arrays of microvilli (Mv) which characterize the normal trophoblast cell surface. Pits or depressions (open arrow head), 0.2–0.4 μm in diameter, in the surface are infrequently observed, perhaps because many are masked by the numerous surface microvilli. × 6554

Figure 4.17 Microvilli (Mv) on the trophoblast surface are reduced in number and more randomly arranged in vinblastine-treated late blastocysts. Short club-like structures (solid arrow head) as well as small blebs (b) are commonly seen. Surface pits (open arrow head) of a diameter indicated in Figure 4.16 are readily evident, possibly because microvilli are less numerous. × 7360

Because of the widespread intracellular drug-induced changes of late blastocysts, scanning electron microscopy of both normal and drug-treated late blastocysts only was performed. Examination of control groups of late blastocysts revealed numerous microvilli characterizing the surface of the trophoblast cells. The structures are of uniform size (0.5–1.0 μm in length; 0.1–0.2 μm diameter). In addition, pit-like depressions are randomly scattered over the normal cell surface (Figure 4.16). In contrast, the surface of trophoblast cells exposed to vinblastine show marked changes in microvillar morphology. These changes include the presence of small bleb-like protuberances as well as aberrantly scattered, short (0.1 μm) club-shaped microvilli (Figure 4.17).

DISCUSSION

The results of attempts to arrive at a better understanding of cellular processes associated with blastocyst development have been presented in this report. An interdisciplinary effort was used in order to: (i) enable correlation of labelling experiments with cell ultrastructure; (ii) facilitate the ordering of predictions based on circumscribed biochemical data; (iii) permit evaluation of the extent of ultrastructural modification resulting from vinblastine:blastocyst interaction; and (iv) facilitate interpretation of ultrastructural analysis in the light of more incisive biochemical knowledge of the synthetic capabilities of blastocysts and of the proposed biosynthetic roles of glucosamine[11,12].

The previous section details the effects of blastocyst development on measurements of total uptake and incorporation of [^{14}C]glucosamine following pulse labelling *in vitro*. The bulk of results pertain to normal blastocyst development which is the framework against which drug-induced changes are judged. Combined labelling and ultrastructural analysis have been completed with early and late blastocysts incubated with vinblastine at a final concentration of 0.1 ng/ml of medium, only. Comments pertaining to possible drug effects are, therefore, restricted to results at this drug titre, which for the purposes of these studies is more informative than conditions leading to complete cellular and blastocyst disruption. Since drug-induced changes were found in late, but not early blastocysts, the discussion is weighted toward observations with late blastocysts.

Quantitative biochemical measurements and radioautographic analysis independently confirm the ability of both early and late blastocysts to utilize [^{14}C]glucosamine for biosynthesis during *in vitro* pulse labelling experiments. With both techniques, the amount of ^{14}C-labelled products accumulated by the embryos is directly proportional to the length of the labelling period. Because of intrinsic procedural differences, the chemical nature of the ^{14}C-labelled products detected by the two procedures is not assumed to be identical. For example, glycolipids may be present in samples used for radioautography, whereas the quantitative precipitation technique primarily measures incorporation of the labelled precursor into macromolecules. Specific cellular physiological factors influencing the extent of labelling after a finite pulse period have been detailed elsewhere[18].

In the presence of vinblastine (0.1 ng/ml) incorporation of [^{14}C]glucosamine is significantly lower than control values for late blastocysts ($p < 0.05$), but not early blastocysts. Similar results are obtained with colchicine, but not lumicolchicine[13] (the biologically inactive analogue). Independent criteria which support the validity of the biochemical measurements with drug-treated late blastocysts include the following: (i) vinblastine-induced cellular and surface ultrastructural modifications of late blastocysts are striking; (ii) subcellular differences in intensity and distribution of grains in radioautographs of normal versus vinblastine-treated late blastocysts are found. This may denote not only impaired processing or distribution of products, but also qualitative differences in the types of molecules being synthesized; and (iii) the 'washed out' appearance of the cytoplasm of vinblastine-treated late blastocysts may indicate that synthesis did not keep abreast with turnover and/or disassembly of organelles, processes which as a consequence of alkaloid treatment may even have been accelerated. Disorganization within the cytoplasm may additionally imply that subcellular compartmentalization of [^{14}C]glucosamine, a factor which could affect incorporation values, had been modified. Functional impairment of the Golgi apparatus itself, a primary site for glycosylation, is possible as a consequence of vinblastine treatment[35].

Interestingly, uptake of [^{14}C]glucosamine (transport and coupled trapping reactions) was not diminished in vinblastine-treated late blastocysts despite striking alterations in the morphology of these embryos. Moreover, it is premature to assume a *direct* connection between diminished incorporation and expression of glycosylated cell surface components in the drug-treated late blastocysts. This association may be difficult to prove since disassembly of cell surface structures and impaired reassembly may result in cellular internalization of surface components. Cell swelling, which can alter morphology, does not seem to be at the root of aberrant surface morphology but has not been unambiguously excluded.

In contrast to results with [^{14}C]glucosamine, both uptake and incorporation of ^{14}C-amino acids remain unchanged for vinblastine-treated late blastocysts. Although protein synthetic activity appears to be normally active, qualitative changes in the types of peptides synthesized and turnover rates of specific peptides may have been modified by alkaloid treatment.

The comparative drug insensitivity of early blastocysts contrasts sharply with findings in late blastocysts. Neither [^{14}C]glucosamine uptake nor incorporation are detectably affected nor is there any evidence of cytoplasmic organelle disruption in early blastocysts treated with 0.1 ng vinblastine/ml.

It is conceivable that early blastocysts may be inaccessible to vinblastine. The zona pellucida has not been unambiguously excluded as a drug permeability barrier[36]. It appears more probable, however, that the trophoblast cell surface membrane which changes with blastocyst development may regulate entry not only of metabolites, but also drugs. A precedent for drug permeability coupled to a cell surface glycoprotein has been reported for Chinese hamster ovary cells[38].

Not only have these studies touched on normal blastocyst development but also on amplification of differences between blastocysts of different gestational age through the use of drug insult. This *modus operandi* appears to provide a feasible and interesting tool in the further analysis of biochemical/functional differences of preimplantation embryos.

ACKNOWLEDGEMENT

These studies were supported in part by grants from NICHD, number HD-06234, and NIH, grant number RR-05384.

References

1. Rugh, R. (1967). *The Mouse: Its Reproduction and Development*, p. 44. (Minneapolis: Burgess Publishing Co.)
2. Bigger, J. D. and Stern, S. (1973). Metabolism of the preimplantation mammalian embryo. *Adv. Reprod. Physiol.*, **6**, 1
3. Papaioannou, V. E., McBurney, M. W. and Gardner, R. L. (1975). Fate of teratocarcinoma cells injected into early mouse embryos. *Nature*, **258**, 70
4. Clementson, C. A. B., Kim, J. K., Mallikarjuneswara, V. R. and Welds, J. H. (1972). The sodium and potassium concentrations in the uterine fluid of the rat at the time of implantation. *J. Endocrinol.*, **54**, 417
5. Holmes, P. V. and Dickson, A. D. (1973). Estrogen-induced surface coat and enzyme changes in the implanting mouse blastocyst. *J. Embryol. Exp. Morphol.*, **29**, 639
6. Nilsson, O., Lindquist, I. and Ronquist, G. (1973). Decreased surface charge of mouse blastocysts at implantation. *Exp. Cell Res.*, **83**, 431
7. Pinsker, M. C. and Mintz, B. (1973). Changes in cell surface glycoproteins before implantation. *Proc. Natl. Acad. Sci. (USA)*, **70**, 1645
8. Enders, A. C. and Schlafke, S. (1974). Surface coats of the mouse blastocyst and uterus during the preimplantation period. *Anat. Rec.*, **180**, 31
9. Nilsson, O., Lindquist, I. and Ronquist, G. (1975). Blastocyst surface charge and implantation in the mouse. *Contraception*, **11**, 441
10. Schlafke, S. and Enders, A. C. (1975). Cellular basis of interaction between trophoblast and uterus at implantation. *Biol. Reprod.*, **12**, 41
11. Hughes, R. C. (1976). *Membrane Glycoproteins: A Review of Structure and Function*, p. 91. (London: Butterworth)
12. Hakomori, S. (1975). Structure and organization of cell surface glycolipids: dependency on cell growth and malignant transformation. *Biochim. Biophys. Acta*, **417**, 55
13. Wilson, L., Bamburg, J. R., Mizel, S. B., Grisham, L. M. and Creswell, K. M. (1974). Interaction of drugs with microtubule proteins. *Fed. Proc.*, **33**, 158
14. Edelman, G. M. and Yahara, I. (1976). Temperature sensitive changes in surface modulating assemblies of fibroblasts transformed by mutants of Rous sarcoma virus. *Proc. Natl. Acad. Sci. (USA)*, **73**, 2047
15. De Brabander, M., De Mey, J., Van de Veire, R., Aerts, F. and Geuens, G. (1977). Microtubules in mammalian cell shape and surface modulation: an alternative hypothesis. *Cell Biol. Int. Reports*, **1**, 453
16. Hynes, R. O. and Destree, A. T. (1978). 10 nm filaments in normal and transformed cells. *Cell*, **13**, 151
17. Brinster, R. L. (1970). *In vitro* cultivation of mammalian ova. *Adv. Biosci.*, **4**, 199
18. Acey, R. A., Hazlett, L. D. and Dabich, D. (1977). Mouse blastocysts pulse labeled with ^{14}C-glucosamine: incorporation and ultrastructural analyses. *Biol. Reprod.*, **16**, 564
19. VanWinkle, L. J. and Dabich, D. (1977). Transport of naturally occurring amino acids and alpha-amino isobutyric acid by normal and diapausing mouse blastocysts. *Biochem. Biophys. Res. Commun.*, **78**, 357
20. Sokal, R. R. and Rohlf, F. J. (1969). *Biometry: The Principles and Practices of Statistics in Biological Research*, p. 387. (San Francisco: W. H. Freeman & Co.)
21. *Ibid*. Ch. 9, p. 204
22. *Ibid*. Ch. 14, p. 404
23. Biggers, J. D. and Borland, R. M. (1976). Physiological aspects of growth and development of the preimplantation embryo. *Annu. Rev. Physiol.*, **38**, 95
24. VanWinkle, L. J. (1975). Factors affecting net incorporation of radioactive amino acids into protein of normal and diapausing mouse blastocysts. Ph.D. Dissertation. Wayne State University, Detroit, Michigan, USA

25. Brinster, R. L., Wiebold, J. L. and Brunner, S. (1976). Protein metabolism in preimplanted mouse ova. *Dev. Biol.*, **51**, 215
26. Acey, R. A. (1977). ^{14}C-glucosamine utilization by preimplantation mouse blastocysts: biochemical and morphological aspects. Ph.D. Dissertation. Wayne State University, Detroit, Michigan, USA
27. Grisolia, S. and Hood, W. (1972). In: E. Kun and S. Grisolia (eds.). *Biochemical Regulatory Mechanisms in Eukaryotic Cells*, pp. 137–204. (New York: Wiley Interscience)
28. Kemp, J. D. and Sutton, D. W. (1971). Protein metabolism in cultured plant tissues. Calculation of an absolute rate of protein synthesis accumulation and degradation in Tobacco Callus *in vivo*. *Biochemistry*, **10**, 81
29. Weitlauf, H. M. (1973). Changes in the protein content of blastocysts from normal and delayed implanting mice. *Anat. Rec.*, **176**, 121
30. Van Blerkom, J., Barton, S. C. and Johnson, M. H. (1976). Molecular differentiation in the preimplantation mouse embryo. *Nature*, **259**, 319
31. Hogan, B. and Tilly, R. (1977). *In vitro* culture and differentiation of normal mouse blastocysts. *Nature*, **265**, 626
32. Dickson, A. D. (1963). Trophoblastic giant cell transformation of mouse blastocysts. *J. Reprod. Fertil.*, **6**, 465
33. Enders, A. C. and Schlafke, S. J. (1965). The fine structure of the blastocyst: some comparative studies. In: G. E. W. Wolstenholme and M. O'Connor (eds.). *Ciba Foundation Symposium. Preimplantation Stages of Pregnancy*, pp. 29–54. (Boston: Little, Brown and Company)
34. McReynolds, H. S. and Hadek, R. (1972). A comparison of the fine structure of late mouse blastocysts developed *in vivo* and *in vitro*. *J. Exp. Zool.*, **182**, 95
35. Minor, P. D. and Roscoe, D. H. (1975). Colchicine resistance in mammalian cell lines. *J. Cell Sci.*, **17**, 381
36. Wilson, I. B. and Stern, M. S. (1975). Organization in the preimplantation embryo. In: M. Balls and A. E. Wild (eds.) *The Early Development of Mammals*. (Cambridge: Cambridge Univ. Press).
37. Juliano, R. L. and Ling, V. (1976). A surface glycoprotein modulating drug permeability in Chinese hamster ovary cell mutants. *Biochim. Biophys. Acta*, **455**, 152

5
Effects of cytochalasin B on preimplantation and early postimplantation mouse embryos *in vitro*

N. H. GRANHOLM

INTRODUCTION
Cytochalasin B and mammalian embryogenesis

The effects of cytochalasins on eukaryotic cells were first reported in 1967[1] following their isolation and characterization[2]. Cytochalasin B (hereafter referred to as CB), a metabolite of the mould *Helminthosporium dermatioideum*, has been reported[1] to modify the *in vitro* behaviour of 'L' strain mouse fibroblasts in the following reversible ways: (1) nuclear extrusion at 1.0–10.0 μg/ml; (2) inhibition of cytokinesis without an apparent influence on karyokinesis at 0.5–1.0 μg/ml; and (3) paralysis of peripheral ruffling membranes (lamellipodia) and cessation of cellular motility at 0.5 μg/ml. In addition, the recognition that CB interferes in some way with 50Å contractile microfilament function[3,4] set the stage for numerous studies dealing with the effects of CB on cells and tissues undergoing primary developmental activities. Examples include neurulation[5–7], cleavage[8–10], salivary gland morphogenesis[11,12], differentiation of epithelial cells in the rat lens[13] and many other morphogenetic processes[14,15]. CB has also been shown to modify various cellular processes including embryonic axon elongation[16], attachment of Ehrlich ascites cells to glass and plastic substrata[17], intercellular adhesion between gastrula cells of *Rana pipiens*[18], endocytosis in polymorphonuclear leukocytes[19], aggregation of dissociated embryonic chick cells[20], cell secretion[21], and numerous aspects of *in vitro* cell motility[22].

Cytochalasins have also been used to investigate various aspects of mammalian embryogenesis. Areas of interest include: (1) the developmental potential of CB-induced parthenogenetically-derived embryos[23,24]; (2) effects of CB-induced polyploidy on developing embryos[25–28]; (3) investigation of pseudocleavage[29] resulting from the action of CB on the cortical contractile apparatus in mouse oocytes; and (4) effects of CB on the morula-to-blastocyst transformation[30, 31] and both short- and long-term effects on early postimplantation embryogenesis[32–34]. A brief review of the current literature in each of these areas will now be presented.

The causes, frequency, and effects of polyploidy on developing mammalian embryos and fetuses are of considerable interest. The frequency of spontaneous polyploidy in *Homo sapiens* is unknown, but at least one tetraploid infant has been born and observed[35]. According to McLaren[36], the early literature on experimental induction of haploidy and tetraploidy has been reviewed[37, 38]; mice of the 'silver' strain have a relatively high incidence of triploid embryos which survive until blastocyst stages[38]. Of the various techniques available for inducing polyploidy in mammalian oocytes, mature ova, and/or diploid embryos[36, 39], procedures employing CB, developed independently by Drs M. H. L. Snow and A. K. Tarkowski, proved to be superior to previous methods[27, 28, 40]. In both methods, tetraploidy is accomplished by a 10.0 μg/ml CB-induced suppression of the second cleavage division *in vitro*; tetraploid embryos are then transferred to pseudopregnant recipients for subsequent development.

According to the protocol of Dr Snow[40], two-cell Q strain mouse embryos are treated for 12–13 h with 10.0 μg/ml CB *in vitro*; 40–75% of the original two-cell embryos produce blastocysts. Upon transfer of 846 presumed tetraploid blastocysts to primed recipients, 31.8% (269/846) implanted, approximately 17% showed postimplantation development, and 0.5% (4/846) developed to term[27]. Three of the four term fetuses were shown to be tetraploids, indicating that CB-induced tetraploidy is compatible with embryogenesis and survival at least until parturition. Interestingly, increased cell size and decreased cell numbers rather than genetic factors were reported to be the cause of developmental aberrations in tetraploid fetuses[27].

Dr Tarkowski's procedure is somewhat different both in terms of treatment duration and age of embryos at transfer. Upon treating F_1 (CBA × C57/BL) × A two-cell mouse embryos at second cleavage with 10 μg/ml CB for 3.0–8.5 (average 4.5 h) pulses, 50% became tetraploid and 20% were found to be diploid/tetraploid mosaics; 90% (9/10) of the transferred presumed tetraploid embryos survived the first three days of postimplantation development[28]. However, the embryonic portion of the eight-day egg-cylinder is retarded[28], signalling the demise of tetraploid embryos.

Preimplantation rabbit embryos when treated with 2.5 μg/ml CB for 24–26 h and transferred to pseudopregnant recipients were able to undergo only limited postimplantation development[25, 41]. Of 105 transferred preimplantation stages, nine or approximately 8% implanted on day 9 or 10; by days 28 or 29 no viable fetuses were recovered.

In the experiments discussed above CB, because of its inhibitory action on cytokinesis, was used to assess the effects of polyploidy on mammalian embryogenesis. However, CB is also utilized to investigate 50Å microfilament function in regions of specialized microfilamentous bundle polymerization like the contractile ring[7, 9, 28] and within the cortical contractile apparatus of mouse oocytes[29, 42]. Within the past few years we have come to understand biological membranes as fluid phospholipid bilayers possessing intercalated, laterally mobile proteins (fluid mosaic model)[43]. Patching, capping, and other surface moiety redistribution phenomena further illustrate dynamic aspects of biological membranes[44]. Evidence of surface modulation and transmembrane control[45] further supports the notion that factors in the cell cortex can influence species, quantity, and distribution of cell surface macromolecules and vice versa. Interestingly, components that may control membrane receptor anchoring include 250Å cytoplasmic microtubules and 50Å microfilaments of the cell cortex[45].

Recent studies by Wassarman and coworkers[29] have probed the relationship between certain surface manifestations of mouse oocytes and CB-induced cortical changes. A low percentage (10–20%) of mouse oocytes when cultured in a medium containing 5 μg/ml CB will undergo a 'pseudocleavage' resulting in two complete spherical components. One component contains the nucleus; the other is anucleate. Surprisingly, all surface microvilli are restricted to the anucleate component. This finding is reminiscent of the observation that surfaces of unfertilized mouse ova possess marked mosaicism with regard to concanavalin A receptor distribution[46]; this asymmetry, a deficiency in concanavalin A receptors overlying the metaphase plate, seems to be influenced by a cortical microfilamentous system. Moreover, with regard to both lectin receptor sites and embryo associated antigenic determinants (H-2 and non-H-2), mouse oocytes and preimplantation embryos appear to possess stage specific surface moieties[47-49]. The investigation of pseudocleavage[29] and studies on the CB-induced redistribution of microfilaments and cell surface components following CB treatments[50] certainly suggest that cortical cytoskeletal elements and cell surface macromolecular activities are coupled and intimately related[45].

Objectives of the present study

In our laboratory we have investigated the effects of cytochalasin B on pre- and postimplantation mouse embryos[30-33]. Objectives of our studies on preimplantation embryos include: (1) identification of ways in which CB alters cellular relationships during the morula-to-blastocyst transformation; (2) acquisition of dose–response data over the cavitation interval using blastocyst formation as the quantal response criterion[51]; and (3) testing of CB as an agent to phenocopy symptoms of the t^{w32}/t^{w32} (Brachyury or T-complex) syndrome. Following these studies, an analysis of latent effects of CB on postblastocyst embryos was conducted[32,33]. Primary objectives of postblastocyst studies were to determine latent effects on postblastocyst embryos of CB pulses at eight-cell stages, to test for differential susceptibility of ICM or trophoblast cell development to CB as reported for certain metabolic inhibitors, and to determine the extent of CB reversibility on long term growth and development of postblastocyst embryos *in vitro*.

MATERIALS AND METHODS
Effects of cytochalasin B on preimplantation embryos

Embryos obtained from random bred ICR stock (Sasco, Inc., Omaha, Nebraska) were recovered[52] at four- and eight-cell preimplantation stages and cultured in 10×35 mm Falcon plastic dishes containing 2.0 ml of Brinster's BMOC-3 (GIBCO) and incubated at 37 °C in an atmosphere of 5% carbon dioxide in air. The overall adequacy of culture conditions is reflected in Table 5.1; by 92 hours *post coitum* (h p.c.) 90.8% of cultured embryos had progressed successfully to blastocyst stages. (See Granholm and Brenner[30] for embryo staging criteria and extensive information on materials and methods).

Table 5.1 Developmental staging in vitro of ICR × ICR control embryos over the 60–92 h p.c. interval

	Number of embryos and percent of embryos per developmental stage per hours post coitum (Hours *post coitum*)				
Stages	60–62	66–68	78–80	84–86	90–92
Four-cell	10 5.7%				
Eight-cell	72 41.4%	10 12.2%	4 3.8%	2 1.7%	
Sixteen-cell	60 34.5%	33 40.2%	3 2.8%	2 1.7%	
Morula	32 18.4%	39 47.6%	64 61.0%	45 37.8%	8 9.2%
EB			24 22.8%	42 35.3%	11 12.6%
MB			5 4.8%	11 9.2%	34 39.1%
DB			5 4.8%	17 14.3%	34 39.1%
Total embryos	174	82	105	119	87

Preliminary studies using 1.0 and 10.0 μg/ml CB allowed a suitable log dose range and a quantal response (all or none) criterion[51], i.e. formation of a blastocyst during the 24 h CB treatment interval (68–92 h p.c.). Randomly chosen groups of 12 embryos (eight-cell stage) were cultured in 4 μl/ml

Table 5.2 Dose–response effect of cytochalasin B on blastocyst formation

	Total embryos out of twelve which formed blastocysts during 24 hour cytochalasin B treatment from 68–92 hours post coitum			
Dose	I	II	III	IV
Control (4 μl/ml DMSO)	10 (12)*	11 (10)	11 (11)	11 (11)
0.25 μg/ml	12 (12)	12 (12)	10 (10)	12 (11)
0.5 μg/ml	6 (12)	8 (12)	10 (12)	10 (12)
1.0 μg/ml	5 (11)	2 (12)	0 (12)	6 (12)
2.0 μg/ml	2 (10)	3 (11)	1 (10)	1 (12)
4.0 μg/ml	0 (11)	0 (11)	0 (12)	0 (12)

* The number in parentheses represents the number of embryos per group of 12 which had transformed into blastocysts after 10 h in Eagle's MEM with 10% fetal calf serum following the CB treatment

DMSO (control) and 0.25, 0.5, 1.0, 2.0, and 4.0 µg/ml CB in BMOC-3, and the number of successful cavitating embryos in each group were recorded (Table 5.2). At 92 h p.c. (following CB treatment) each group of 12 embryos was rinsed with control BMOC-3 and cultured in Eagle's MEM with 10% fetal calf serum for an additional 10 h to assess recovery from CB. The total number of embryos per group of 12 which successfully cavitated upon recovery is shown in the parentheses of Table 5.2.

Effects of cytochalasin B on postblastocyst embryos

Two independent sets of experiments were conducted. In the first set, definitive blastocysts (controls) were cultured in Eagle's MEM containing 10% fetal calf serum (GIBCO). Following their attachment to Corning glass coverslips and Falcon plastic culture dishes, they were treated for 30–120 min with either 1.0 or 10.0 µg/ml CB, rinsed, and cultured in control medium over a recovery period. Both still and time-lapse photographs were taken; cinemicrographic sequences were analysed using a Lafayette projector. Data on CB-induced effects of outgrowth size are presented in Table 5.3.

Table 5.3 Effects of 10.0 µg/ml cytochalasin B on areas of trophoblast outgrowths

		Areas (expressed in square inches at × 205 magnification)			
		70–75 minutes CB treated (10.0 µg/ml)		21 hours post CB at 175 h p.c.	
Trophoblast outgrowth No.	Controls 153 h p.c.	Area	Area after contraction (% control)	Area	Area after expansion (% control)
1	2.52	2.17	86.1	2.68	106.3
2	3.61	2.66	73.7	6.32	175.1
3	5.16	4.44	86.0	5.71	110.6
4	3.30	2.89	87.6	4.18	126.7
Average	3.65	3.04	83.4	4.72	129.7

In the second set of experiments, eight-cell ICR embryos were treated with 4.0 µg/ml CB for 6, 12, 18 and 24 h intervals, rinsed in control media, and cultured for extended durations in Eagle's MEM/10% fetal calf serum. In long-term cultures media were replaced (exchanged) every 24 h. Following zona escape and expanded blastocyst attachment at 120–130 h p.c., all groups were scored for attachment success, outgrowth expansion, and differentiation of ICM and trophoblast components at selected time periods. Quantitative data were acquired on the following postblastocyst parameters: (1) attachment; (2) outgrowth size; (3) ICM size; and (4) extent of peripheral hyaloplasmic fan per outgrowth. Photographs (35 mm) at 50× magnification were taken at each of the selected time periods.

Contact prints of each outgrowth were analysed for areas of outgrowth, ICM, and hyaloplasmic fan. Areas were determined by superimposing a transparent calibrated grid containing evenly spaced points over individual

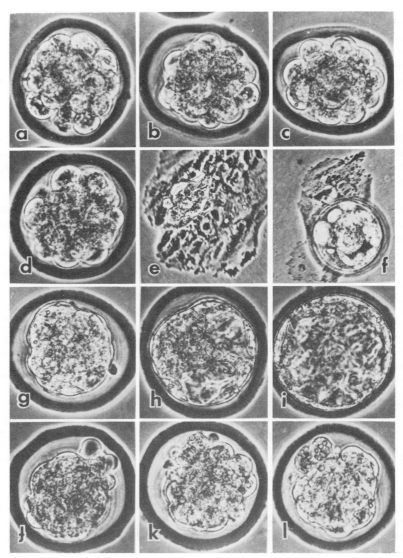

Figure 5.1 a–d, Early blastocysts treated with 10.0 μg/ml CB for 14 h from 78 to 92 h p.c. (× 343); e, f, Trophoblast outgrowths at 120 h p.c. following 10.0 μg/ml CB treatment over the 92 to 102 h p.c. interval (× 343); g–i, Morula, mid-blastocyst, and definitive blastocyst respectively at 90 h p.c. after treatment with 1.0 μg/ml CB over the 11 h interval between 79 to 90 h p.c. (× 343); j–l, Embryos at 101 h p.c. after treatment with 1.0 μg/ml CB over the 22 h interval between 79 to 101 h p.c. (× 343)

contact prints and counting points falling on or within the perimeter of ICMs, fans, and outgrowths. This method determines absolute area; absolute area = nd^2, where d is the interdot distance[53]. Finally, data were statistically analysed by chi square (χ^2) and analysis of variance (ANOVA).

RESULTS
Effects of CB on preimplantation embryos
General effects
Effects of 10.0 μg/ml CB on preimplantation embryos were striking. Following a 14 h (78–92 h p.c.), 10.0 μg/ml CB treatment, early blastocysts transformed into 'morulae' (Figure 5.1a–d). In addition to taking on the morula appearance it was obvious that embryos became more transparent (less translucent); thus CB causes an overall change in embryo opacity. Following CB treatment (10.0 μg/ml for 10 h, 92–102 h p.c.) of blastocysts, embryos were rinsed and cultured in Eagle's MEM/10% FCS; three of six contracted blastocysts expanded, hatched from their zonae, and attached forming outgrowths (Figure 5.1e, f). Retention of zona and presence of abnormally large and heterogeneously-appearing blastomeres suggest significant abnormalities.

Following 1.0 μg/ml CB treatment of three morulae and six early blastocysts from 79–90 h p.c., embryos at 90 h p.c. were staged as three morulae, six early and midblastocysts, and one definitive blastocyst (Figure 5.1g–i). At 101 h p.c., after a 22 h, 1.0 μg/ml CB pulse, some of the treated embryos were still arrested resembling morulae (Figure 5.1j–l). Upon rinsing and culture in control media, arrested embryos formed blastocysts by 108 h p.c. (Figure 5.2a–c). During recovery from CB treatments, embryos assumed smoother peripheral profiles and became increasingly granular or more translucent in appearance.

Dose–response relationships
In order to define dose–response relationships of CB on preimplantation mouse development, groups of 12 embryos (eight-cell stage) were treated for 24 h (68–92 h p.c.) and scored at 92 h p.c. for their ability to undergo cavitation and form blastocysts. Four sets of data were obtained for each of the five concentrations tested plus the control (4.0 μl/ml DMSO) group (Table 5.2). Differences between CB concentrations were highly significant (ANOVA, $p<0.01$). After converting data in Table 5.2 to percentage of nondevelopers and converting those to probits, a probit-response–log dose line was fitted to the data[54] (Figure 5.3). The equation for the CB dose-response line is $y=-2.965+2.779x$, where x represents CB concentration and y, the expected probit response. The dose–response line was highly significant. A chi square goodness of fit test was not significant, indicating that our data were adequately represented by the dose–response line. The ED_{50}[55] (median effective CB dose) or concentration of CB required to inhibit 50% of the treated embryos was estimated to be 0.73 μg/ml with 5% confidence limits of 0.58 to 0.89 μg/ml. Finally, the concentration of CB needed for a 95% inhibition was estimated to be 2.87 μg/ml with 5% confidence limits of 2.16 to 4.43 μg/ml.

At a concentration of 4.0 μg/ml, CB inhibits cytokinesis, and embryos

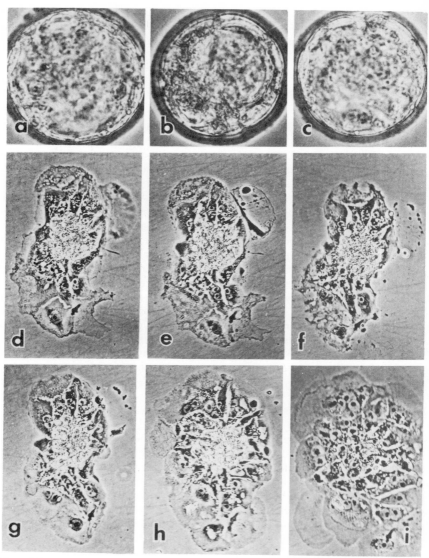

Figure 5.2 a–c, Blastocysts at 108 h p.c. following replacement of 1.0 μg/ml CB solution with Eagle's MEM/10% FCS at 102 h p.c. The embryos had been recovering for 6 h following a 1.0 μg/ml CB treatment for 22 h over the 79 to 101 h p.c. interval (× 343); d, Control trophoblast outgrowth at 153 h p.c. (× 122.5); e, Same outgrowth after 10.0 μg/ml CB 5 min, 153 h p.c. (× 122.5); f, Same outgrowth after 10.0 μg/ml CB, 72 min, 154 h p.c. (× 122.5); g, Same outgrowth after 31 min recovery in Eagle's MEM/10% FCS, 155 h p.c. (× 122.5); h, Same outgrowth after 8 h recovery, 163 h p.c. (× 122.5); i, Same outgrowth after 20 h recovery, 175 h p.c. (× 122.5)

EFFECTS OF CB ON EARLY MOUSE EMBRYOS

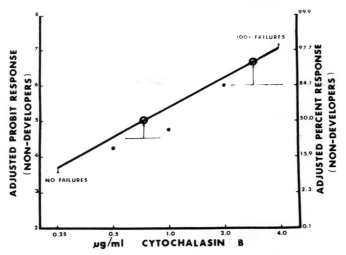

Figure 5.3 The log dose effect of cytochalasin B on the morula-to-blastocyst transformation over the 68–92 h treatment period. The doses required to inhibit blastocyst formation in 50 and 95% of embryos are indicated on the graph with their 5% fiducial limits.

retain the cell number they possessed when initially subjected to the drug. Eight-cell embryos when treated with 4.0 μg/ml CB at 68 h p.c. remained in the eight-cell stage. Eight-cell stages which had previously compacted[56,57] underwent a reverse compaction process[58] characterized by apparent changes in embryo transparency and blastomere shape; CB induced obvious modifications in the extent of cell-to-cell contacts.

Recovery from the effects of cytochalasin B

The ability of embryos to resume development following their release from CB-induced developmental blocks was quite striking. For example, in spite of the 100% inhibition of blastocyst formation at the highest CB concentration (4.0 μg/ml) in our dose–response trials, 95.8% (46/48) formed blastocysts during the 10 h interval after rinsing and culture in Eagle's MEM/10% FCS (Table 5.2, data in parentheses). Careful morphological analyses of some 'recovered' blastocysts revealed the presence of partitioning or reticulation of blastocoelic spaces by elongate cytoplasmic strands, thickening of trophoblast epithelia (Figure 5.2b), several ICM foci or clusters (ectopic ICMs), and abnormal trophoblast surface bulges.

Effects of cytochalasin B on postblastocyst embryos

Outgrowth phenomena in control embryos

When flushed from oviducts at 60 h p.c., ICR embryos are generally at the four-cell or various eight-cell stages, i.e. uncompacted, partially compacted, and/or fully compacted. During the next 24 h ICR embryos develop to morulae and early blastocysts (Table 5.1). By 120 h p.c. blastocysts greatly expand, shed zonae pellucidae, and begin attachment to suitable substrata.

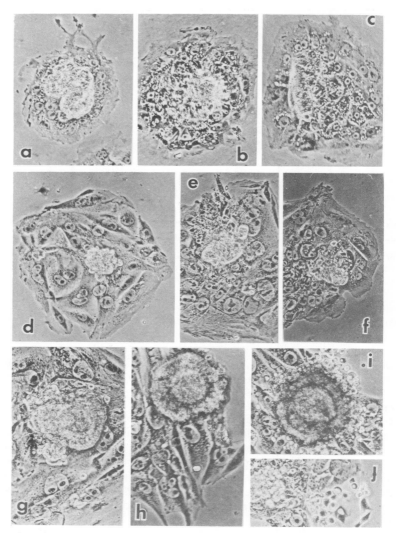

Figure 5.4 Control outgrowths. a–c, Early outgrowth stages showing peripheral trophoblast and central ICM. Note monolayer-like trophoblast epithelium containing small nuclei and clear peripheral hyaloplasmic regions (especially Figure 5.4a), 161 h p.c. ($\times 119$); d–f, Later outgrowth stages possessing well-defined ICMs. The trophoblast layer contains giant cells with nuclei often measuring 10–50 μm in diameter, 199 h p.c. ($\times 119$); g, Well-developed egg cylinder growing upwards into the medium. Nuclei of giant trophoblast cells are obvious. One can just detect an outline of the proamnionic cavity at the lower centre of this egg cylinder, 224 h p.c. ($\times 119$); h, Outer entodermal cells, inner embryonic ectodermal cells, and the internal proamniotic cavity can be observed in this well-differentiated egg cylinder, 248 h p.c. ($\times 119$); i, Well-differentiated egg cylinder undergoing peripheral entoderm dissociation, 275 h p.c. ($\times 119$); j, Entodermal cells plating out on a plastic substratum following peripheral entoderm cell dissociation similar to that shown in Figure 5.2i. Note fibroblast-like conformation of attached entoderm cells, 321 h p.c. ($\times 119$)

Following attachment, postimplantation development involves differentiation of the inner cell mass to form an egg cylinder complete with proximal and distal entoderm plus continued differentiation and growth of the trophoblast component.

Initial substratum attachment of expanded blastocysts appears to occur via lateral and/or abembryonal trophoblast cells. Once attached, the trophoblast layer ruptures and spreads along the substratum. As a result of this developmental feature unique to the *in vitro* situation, the ICM becomes a superficial clump of cells in direct contact with culture medium. Within five to ten hours of initial embryo attachment, trophoblast cells spread peripherally (Figure 5.4a–c); in these early outgrowth stages (161 h p.c.) ICMs initially appear to flatten on the upper surface of trophoblast monolayers. ICMs either spread over a large portion of the trophoblast (Figure 5.4a, b), remain as discrete compact components (Figure 5.4c, d), or disappear. Trophoblast cells are easily recognized as exceedingly flat, pavement-like epithelia possessing thin hyaloplasmic fans and ruffling membranes at their margins. At 199 h p.c., some outgrowths possess discrete and well defined egg cylinders which project upward into the medium (Figure 5.4d–f); trophoblast cells increase in size, become vacuolate, contain giant nuclei (Figure 5.4e), and exhibit perinuclear granulation (Figure 5.4d). Outgrowths pictured in Figure 5.4g–i possess well-differentiated egg cylinders. In Figure 5.4g one can detect concentric rings within that 224 h p.c. egg cylinder. The outgrowth pictured in Figure 5.4h (248 h p.c.) possessed well-differentiated layers composed of outer (proximal) entoderm and inner embryonic ectoderm.

The extent to which egg cylinder differentiation in control embryos resembles normal development is questionable. Sloughed entodermal cells of egg cylinders are observed in Figures 5.4g and 5.4i; peripheral entoderm cells are attached to the Falcon plastic substratum (Figure 5.4i). Thus, at least one abnormality, peripheral dissociation of egg cylinder entodermal cells, occurs under the present culture conditions.

Upon gentle swirling of culture dishes, 224 h p.c. egg cylinders were seen to sway back and forth in the medium. Heights of 13 egg cylinders at 224 h p.c. were measured using fine focus microscope setting focussed on the top cell layer of egg cylinders and on the trophoblast monolayer. The difference of the two measurements was considered to approximate egg cylinder heights. Heights ranged from 21 μm to 117 μm with an average height of 54 μm.

Treatment of control outgrowths with cytochalasin B
Upon attachment, trophoblast cells of outgrowths spread uniformly, and there did not appear to be any net directional movement of outgrowths. Clear, hyaloplasmic fan-like areas were seen to exist. Interestingly, peripheral hyaloplasmic regions, when analysed with time-lapse cinemicrography, exhibited ruffling membrane activity. The time required for complete propagation and retraction cycles of ruffling membrane activity (n = 6) ranged from 8 to 14 min per cycle. CB at 1.0 or 10.0 μg/ml caused immediate cessation of ruffling membrane activity coupled with retraction of hyaloplasmic fans at 10.0 but not 1.0 μg/ml CB. Upon rinsing and culture in control media, fans and ruffling membrane activity were restored (Figure 5.2d–i). Data in Table 5.3 illustrate

changes in outgrowth areas in the presence of 10.0 μg/ml CB. Apparently most of the outgrowth contraction and expansion (average values, Table 5.3) were accomplished by the loss and regeneration of peripheral fan regions. Following CB treatments, the production of excess peripheral fan areas was greatly enhanced (Figure 5.2i and Table 5.3).

Latent effects of cytochalasin B on outgrowths

Latent effects of varying durations of 4.0 μg/ml CB on postblastocyst morphogenesis were analysed in the following manner. Eight-cell ICR embryos were treated at 68 h p.c. for 6, 12, 18, and 24 h pulses, rinsed, cultured in MEM/10% FCS for periods of up to 240 h p.c. (10 days p.c.), and scored for a number of parameters over five time intervals. These include attachment of

Table 5.4 Attachment of embryos following cytochalasin B treatments

Duration of CB treatment (hours)	Percent and number of embryos attached (Hours *post coitum*)		
	135	153	200
DMSO control (4.0 μl/ml DMSO)	69.6 (128/184)	76.4 (133/174)	68.2 (90/132)
6	63.6 (28/44)	53.7 (22/41)	43.9 (18/41)
12	52.8 (47/89)	77.4 (65/84)	55.4 (31/56)
18	25.4 (18/71)	41.7 (30/72)	27.1 (13/48)
24	12.1 (13/107)	32.9 (28/85)	25.8 (23/89)

Table 5.5 Effect of varying durations of 4.0 μg/ml CB at 68 h p.c. on ICM morphogenesis in 120–240 p.c. outgrowths

	Percent and number of outgrowths possessing ICMs over five time intervals (Hours *post coitum*)						
	120–143	144–167	168–191	192–215	216–239	Average*	Weigh. averag†
DMSO Control	93.9% (31/33)	91.2% (73/80)	86.1% (31/36)	80.0% (60/75)	83.6% (56/67)	86.8	86.2
6 h CB	55.6% (10/18)	80.0% (16/20)	62.5% (5/8)	50.0% (8/16)	71.4% (5/7)	63.9	63.8
12 h CB	71.4% (15/21)	64.1% (25/39)	52.4% (11/21)	32.4% (11/34)	15.8% (3/19)	47.2	48.5
18 h CB	57.1% (4/7)	38.1% (8/21)	60.0% (6/10)	36.8% (7/19)	20.0% (2/10)	42.4	40.3
24 h CB	66.7% (2/3)	31.6% (6/19)	26.7% (4/15)	20.0% (4/20)	8.3% (1/12)	30.7	24.6

* Average over the five time intervals   † Weighted average over the five time intervals

Table 5.6 Outgrowth area* versus time in control groups

Experiment	120–143		144–167		168–191		192–215		216–239		ANOVA†	
	n	Area ± SE	n	Area ± SE	n	Area ± SE	n	Area ± SE	n	Area ± SE	df n/d	F no.
I			15	67.7 ± 4.7‡	4	97.8 ± 8.0	16	109.0 ± 4.0	13	125.7 ± 4.5	3/44,	27.52§
II			13	42.8 ± 3.5	11	66.4 ± 2.8	13	87.8 ± 6.2	16	95.5 ± 7.5	4/57,	22.54§
III	9	16.4 ± 3.0	30	69.3 ± 3.1			25	66.9 ± 7.0			2/76,	5.58§
IV	24	48.1 ± 3.4	22	78.4 ± 5.8	21	67.0 ± 3.9	21	88.1 ± 7.4	21	95.8 ± 9.9	3/81,	2.94‖
ANOVA†: df n/d, F no.	1/31, 27.25‖		3/76, 8.50§		2/33, 6.48§		3/71, 6.35§		2/47, 3.34‖			

*Areas were measured by superimposing a grid of uniformly spaced points (2.0 mm spacing) over $50 \times$ contact photographs of outgrowths; point counts are presented in the table. Areas in square units can be determined by using the expression $A = nd^2$ where n is the number of points and d the distance between points
† Analysis of variance with numerator and denominator degrees of freedom (df n/d) and F number (F no.)
‡ Data presented in means ± standard errors
§ $p < 0.01$
‖ $p < 0.05$

Table 5.7 Outgrowth area* versus 4.0 µg/ml CB duration – experiment II

						Observation times (h p.c.)					
		120–143		144–167		168–191		192–215		216–239	
Treatments	n	Area ± SE	n	Area ± SE	n	Area ± SE	n	Area ± SE	n	Area ± S
Control	9	16.4 ± 3.0†	13	42.8 ± 3.5	11	66.4 ± 2.8	13	87.8 ± 6.2	16	95.9 ± 7.
12 hour CB	3	22.0 ± 1.9	7	39.7 ± 4.4	6	53.0 ± 12.4	6	61.0 ± 4.9	6	56.0 ± 1(
18 hour CB	3	22.7 ± 2.4	7	42.0 ± 2.2	6	54.3 ± 6.4	6	47.3 ± 7.3	6	55.7 ± 7.
24 hour CB	1	20.0 ± 0	2	27.0 ± 7.8	2	34.0 ± 14.2	5	31.2 ± 6.4	2	34.0 ± 0
ANOVA‡ df n/d, F no.		3/12, 0.06		3/25, 1.08		3/21, 3.60§		3/26, 12.14‖		3/26, 6.12	

* Areas were measured by superimposing a grid of uniformly spaced points (2.0 mm spacing) over 50× contact phot graphs of outgrowths; point counts are presented in the table. Areas in square units can be determined by using t expression $A = nd^2$ where n is the number of points and d the distance between points
† Data presented in means ± standard errors
‡ Analysis of variance with numerator and denominator degrees of freedom (df n/d) and F number (F no.)
§ $p < 0.05$
‖ $p < 0.01$

Table 5.8 ICM area versus 4.0 µg/ml CB treatments*

	Number and area† of ICMs at five observation intervals				
Treatments	120–143	144–167	168–191	192–215	216–239
Control‡	9 6.0 ± 0.9	13 5.2 ± 0.8	11 4.4 ± 0.8	13 6.8 ± 1.1	16 8.1 ± 1.2
12 h CB	3 8.0 ± 1.0	7 4.0 ± 0.7	6 2.7 ± 1.2	6 0	6 1.3 ± 1.2
18 h CB	3 2.0 ± 1.6	7 2.3 ± 1.1	6 2.7 ± 1.2	6 3.7 ± 1.6	6 2.7 ± 1.5
24 h CB	1 0	2 3.0 ± 2.1	2 0	5 2.4 ± 1.3	2 0
ANOVA§	3/12 3.35	3/25 1.57	3/21 1.88	3/26 5.36‖	3/26 7.45‖

* One of four replications
† Areas were determined using a point grid system (see Methods) and are presented in means ± standard errors
‡ 4.0 µl/ml DMSO
§ Analysis of variance with numerator and denominator degrees of freedom and F number
‖ $p < 0.01$

embryos to the plastic substratum (Table 5.4), outgrowths possessing ICM components (Table 5.5), outgrowth area (Tables 5.6 and 5.7), ICM area (Table 5.8), and extent of peripheral hyaloplasmic fan production (Table 5.9).

Embryo attachment data are presented in Table 5.4. Highly significant differences (χ^2, $p < 0.01$) were observed between groups at 135, 153, and 200 h p.c. An inverse relationship is observed between CB pulse duration and ability of embryos to attach.

Percentage and number of outgrowths possessing ICMs over the five time intervals are shown in Table 5.5. The average and weighted average figures

Table 5.9 Effect of varying durations of 4.0 μg/ml CB at 68 h p.c. on hyaloplasmic fan production in 120–240 h p.c. outgrowths

Treatment		Percent of outgrowths possessing abnormally large hyaloplasmic fan areas* (% Ab Out) and percent of total outgrowth area composed of abnormal hyaloplasmic fans (% Area) over five time intervals				
		120–143	144–167	168–191	192–215	216–239
DMSO control	% Ab Out	0	0	0	0	0
	% Area	0	0	0	0	0
6 h CB	% Ab Out	0	0	0	0	0
	% Area	0	0	0	0	0
12 h CB	% Ab Out	0	8/39 (20.5)	3/21 (14.3)	1/34 (2.9)	0
	% Area	0	36.1	47.6	24.6	0
18 h CB	% Ab Out	4/7 (57.1)	9/21 (42.8)	3/10 (30.0)	2/19 (10.5)	0
	% Area	70.0	76.6	41.3	45.4	0
24 h CB	% Ab Out	2/3 (66.7)	15/19 (78.9)	5/15 (33.3)	5/20 (25.5)	3/12 (25.0)
	% Area	65.1	45.5	48.3	38.0	77.4

*Areas were measured using a dot count technique described in the Methods

(last two columns, Table 5.5) of outgrowths possessing ICMs in each of the five treatment groups clearly show that CB duration and presence of ICM components in outgrowths are inversely related.

Four replications of the effects of CB on outgrowth size were conducted. Data in Table 5.6 summarize the control situation in each of the four replications. Control outgrowths increased significantly in area over time (last column, Table 5.6). Significant differences also occurred between control data at each of the five time intervals (bottom row, Table 5.6); thus data from each of the four replications could not be pooled. However, data in Table 5.7 represent one of the four replications and serve to illustrate the effects of CB on outgrowth area. Conclusions drawn from Table 5.7 include: (1) control outgrowths increased consistently in area over the five time intervals (ANOVA, $F = 22.54$, $p < 0.01$); (2) differences in outgrowth area (bottom row, Table 5.7) occurred as a result of increases in CB duration; (3) generally, decreases in outgrowth area became more pronounced with time; and (4) as opposed to control and 6 h CB treatment groups, outgrowths within 12, 18, and 24 h CB groups tend to grow out to their maximum areas by 168–191 h p.c. As seen with outgrowth areas, ICM areas were also inversely affected by CB duration (Table 5.8).

Some embryos within 12, 18, and 24 h treatment groups possessed abnormally extensive peripheral hyaloplasmic fan regions. Data in Table 5.9 indicate a direct relationship between CB duration and extent of abnormal hyaloplasmic fan production. The percent of outgrowths with abnormally large hyaloplasmic regions was greatest during the early time periods and then declined over succeeding intervals.

DISCUSSION
Mode of action of cytochalasin B

Prior to making interpretations based on the present data, it is useful to briefly review concepts dealing with precise modes of action of cytochalasin B. CB may act: (1) directly on 50Å microfilaments and/or their subunits[14]; (2) indirectly on 50Å microfilament function by disrupting insertion of microfilaments into the plasma membrane[59]; and/or (3) directly on membranes[60], especially membrane transport[29,61]. Data which support a direct action of CB on membrane function such as membrane-bound transport systems are accumulating[61]. In addition, CB has been shown to interact directly with purified actin and/or myosin[62] suggesting a causal relationship between CB and an impairment of the cell's cytoskeletal contractile apparatus. However, conflicting data on the interactions of CB and contractile proteins make this particular research area a highly controversial one[14]. Other effects of CB include: (1) inhibitory effects on sugar transport[63]; (2) depression of mucopolysaccharide synthesis[61]; (3) interaction of ^{3}H-labelled CB with lipid monolayers[64]; (4) reduction of electrophoretic mobility of *Rana pipiens* gastrula cells[18]; other aspects have recently been reviewed[10,14,29]. Two additional reports dealing with effects of cytochalasin D on established cell lines[65,66] are also valuable reviews.

Effects of cytochalasin B on preimplantation embryos

CB at 10.0 and 4.0 µg/ml induced complete but reversible developmental arrest of preimplantation mouse embryos. At these concentrations cytokinesis is arrested, while nuclear division continues, resulting in polyploidy[28,40,67]. Other CB-induced changes include the transformation of early blastocysts to morula-like embryos having distinct and recognizable blastomeres (Figure 5.1a–d), a reverse compaction process of previously compacted eight-cell stages[58], and the readily observable modification of embryo opacity from a translucent to a more transparent condition in the presence of CB. These former results show that CB has a rather dramatic effect on cell shape in preimplantation embryos. Blastomeres of early blastocysts as well as eight-cell stages assume more rounded forms in the presence of CB. Cell–cell contact[31] and perhaps other junctional phenomena undergo major modifications as a result of CB-induced blastomere shape changes.

Ultrastructural studies[31] of eight-cell mouse embryos following 12 and 22 h pulses of 4.0 µg/ml CB show that CB induces changes in cell contacts, composition of cortical cytoplasm, blastomere surface contour, integrity of cell cortices, number of nuclei per blastomere, and organization of cytoplasmic organelles. Similar results have been reported[68] after 5 minutes to 12 h, 10.0 µg/ml CB treatments; irregular spacing of microvilli, alterations in blastomere surface contour, coated pits (local plasma membrane invaginations), modified cell junctions, segregation of cortical from internal cytoplasm, and increased deposition of crystalloid inclusions were observed. Interestingly, the above findings may be interpreted as consequences of failure of normal cortical function due either to the direct action of CB on 50 Å contractile microfilaments or to its indirect or secondary effects via microfilament–membrane insertion or other membrane-mediated phenomena.

Dose–response data (Table 5.2, Figure 5.3) indicate a gradual trend of almost no effect on cavitation of preimplantation of embryos when cultured with 0.25 μg/ml CB to a complete developmental block at 4.0 μg/ml. Of course, these dose–response data are based on the ability of embryos to successfully cavitate; by selecting other quantal response criteria[51], different dose–response results would be obtained. Nonetheless, 4.0 μg/ml CB was the lowest concentration that inhibited 100% (48/48) of the embryos tested (Table 5.2); other investigators[28,40,67] have used a similar concentration (10.0 μg/ml CB) to arrest cytokinesis and induce polyploidy in two-cell stages and to investigate compaction in eight-cell embryos[58].

The ability of CB-treated embryos to recover from the treatment was quite extraordinary. Data in parentheses (Table 5.2) show the number of embryos per group of 12 which developed to blastocysts following 10 h of culture in a control medium. In the 4.0 μg/ml CB group 95.8% (46/48) of these long term (24 h) treated embryos underwent cavitation. Morphological analyses of recovered blastocysts from 4.0 μg/ml CB groups revealed several abnormalities including partitioning or reticulation of blastocoelic spaces by elongate cytoplasmic strands, presence of clusters of ICM cells (ectopic ICMs) positioned irregularly around the inner trophoblast epithelium, thickened trophoblast epithelia, and abnormal trophoblast surface bulges. Occurrence of these morphological abnormalities suggests that CB prevents the orderly differentiation and segregation of ICM and trophoblast components, possibly by interfering in specific ways with morphogenetic cell movements[31,65,68].

In some cases 4.0 μg/ml CB treated and recovered blastocysts possessed only 8–16 blastomeres. This particular observation underscores a very intriguing aspect of preimplantation mouse morphogenesis, i.e. the presence of an innate chronological or developmental clock[67,69] which is regulated by some factor(s) other than cell number. Such regulatory candidates include cycles of DNA synthesis, chromosome replications, nucleocytoplasmic ratio, and/or other factors relating to embryo age[67].

The possibility that CB 'phenocopies' certain effects of the t^{w32}-allele is a most provocative consideration. Recent reviews of T-locus developmental genetics[70,71] and reports on specific lethal t alleles[72-75] discuss evidence for the impairment of cell surface and/or cell cortical function in T-locus generated lethalities. For example, t^{w32} homozygotes undergo an aberrant eight-cell compaction response[72] which resembles the CB-induced reverse compaction process[58]. Ultrastructural observations on phenotypically abnormal cells of both T/T and t^9/t^9 embryos[73] are compatible with the concept that T and t^9 gene products alter cell–cell interactions resulting in failures of cell recognition, cell movements, and other requisite morphogenetic processes. In arresting t^9 homozygotes primitive streak cells appear to be immobilized and do not elaborate filopodia; the blunt protuberances that are produced do not possess 50 Å cortical microfilaments as seen in control primitive streak cells[74]. Furthermore, cellular junctions between primitive streak cells of t^9 homozygotes are clearly abnormal[74]. Absence of 50 Å microfilaments[68], modification of cell junctions[30,31,68], and cell recognition failures[32] (ectopic ICMs) are also features common to CB-treated preimplantation mouse embryos. These data suggest that CB does indeed act as a partial phenocopy of t^{w32} and possibly other lethal alleles of the T-locus.

Effects of cytochalasin B on postblastocyst embryos

Treatment of outgrowths with cytochalasin B

Effects of CB on trophoblast expansion and membrane activity of outgrowths (Figure 5.2d–i and Table 5.3) were predictable. CB at both 10.0 and 1.0 μg/ml caused cessation of trophoblast cell ruffling membrane activity followed by contraction (in 10.0 but not 1.0 μg/ml CB) of peripheral hyaloplasmic fan regions. Upon rinsing and control culture of outgrowths, restoration of both fan regions and ruffling membrane activity quickly resumed. Figure 5.2d–i clearly shows that most of the outgrowth contraction and expansion Table 7.3) were accomplished by the loss and regeneration of peripheral hyaloplasmic fan regions. Outgrowths treated with CB, rinsed, and cultured in the control medium (Eagle's MEM/10% FCS) produced abnormally extensive fan regions (Figure 5.2i and Table 5.3). Reasons for this specific effect of CB are unknown, but it would seem that trophoblast outgrowths with their ruffling membrane activity, peripheral hyaloplasmic fan components, and known 'invasive' character would be ideal models for studies on cell motility, contact phenomena and invasiveness.

Latent effects of cytochalasin B pulses at eight-cell stages on postblastocyst morphogenesis

In addition to preimplantation effects, data presented in this paper and elsewhere[32,33] indicate that treatment of 68 h p.c. embryos with 4.0 μg/ml CB for 6 to 24 h also impairs the subsequent attachment of blastocysts to substrata and the development of ICM and trophoblast cells.

Attachment of outgrowths

Data in Table 5.4 reveal significant differences (χ^2, $p < 0.01$) between treatment groups at 135, 153, and 200 h p.c. observation times. In another independent set of data[33], attachment of blastocysts at 130 h p.c. was different between treatment groups (χ^2, $p < 0.01$); percent attachment in each of the groups was 84.3% (59/70), 79.7% (55/69), 67.6% (46/68), 36.2% (25/69), and 23.9% (17/71) for control, 6, 12, 18, and 24 h CB treatment groups respectively. Although control attachment rates were somewhat lower than those described by others[76-78] it appears that CB does affect attachment of embryos in a duration-dependent manner.

CB may exert this latent effect on attachment by: (1) interfering with hatching from the zona pellucida, (2) altering the capability of CB-sensitive trophoblast exploratory processes to make contact and attach to the plastic substratum, and/or (3) decreasing the number of ICM cells below a critical threshold level[79] thereby secondarily inhibiting trophoblast cell attachment and/or other factors. In a related study[80] 58.7% (37/63) of CB-induced tetraploid blastocysts produced outgrowths *in vitro* as compared to 96.7% (29/30) of the controls.

Growth and differentiation of ICM and trophoblast components

Results of this study show that 4.0 μg/ml CB pulses at 68 h p.c. significantly affect the presence of ICM components (Table 5.5), outgrowth area (Tables 5.6 and 5.7), ICM area (Table 5.8), and extent of abnormal hyaloplasmic fans (Table 5.9) in a duration-dependent manner. In another independent study[33], the effects of 6, 12, 18, and 24 h CB durations on numbers of trophoblast cell nuclei at 130, 160, and 190 h p.c. were also shown to be significantly lowered (ANOVA, $p < 0.01$). It thus appears that the approximate degree of latent CB-induced defects can be predictably determined by the duration of 4.0 μg/ml CB treatment. In short, our data reflect a summation or accumulation of defects directly related to CB duration. The degree of ploidy also increases with CB duration[1]. At the time of CB treatment (eight-cell stage), embryos are entering their fourth cleavage division. It is reasonable to assume that 6, 12, 18, and 24 h CB pulses induce increasingly higher levels of ploidy ($4n$, $8n$, perhaps $16n$, and mosaics of different ploidy levels). Accordingly, our duration–response results may be partially explained by increased levels of ploidy. Although tetraploidy has been shown to be compatible in mice embryogenesis at least to parturition[27], octaploidy or $4n/8n$ mosaicism may pose insurmountable physical and genetic problems.

In addition to effects of polyploidy, the precise time interval (68–92 h p.c.) during which embryos were treated with CB may also be related to the incidence of latent defects in outgrowths. The morula-to-blastocyst developmental interval represents a period of intensive biosynthetic activity[36,81,82]. Morulae treated for 24 h with 1×10^{-7} M 5-bromodeoxyuridine were analysed for developmental success at outgrowth stages; results indicated that gene expression (transcription) for ICM and trophoblast components occurs during cavitation[83]. It is also well-documented that paternal and, thus, embryo originated gene products are synthesized during cleavage and morula stages[36,81]; the observed developmental retardation in two- and four-cell stages of C57BL/6J lethal yellow (A^y/A^y) homozygotes[83-85] supports this general model of embryo gene expression (transcription) during early mammalian embryogenesis. Thus CB, although not having a direct inhibitory action on transcription and protein synthesis, may alter the regulation of differential gene expression over the 68–92 h interval by its effects on membrane transport of essential regulatory factors, substrates, etc., by its influence on coordinated activities of the cell surface and cell cortex (transmembrane control)[45], or by other activities. Indeed, it would be most interesting to compare the developmental success of transferred tetraploid embryos produced by suppression of the fourth cleavage division (4.5 h, 10.0 μg/ml CB pulse from 68–72.5 h p.c.) to tetraploids produced under the same conditions at the second cleavage division[27,28,80].

Metabolic inhibitors of RNA or protein synthesis (cordycepin, actinomycin D, and cyclohexamide) selectively inhibit ICM differentiation, while trophoblast development proceeds unaffected[86,87]. Specific concentrations of three other antimetabolites (bromodeoxyuridine, cytosine arabinoside, and colcemid) also cause differential damage to ICM which dies, while trophoblast cells develop normally[88]. Although 4.0 μg/ml CB does indeed induce specific latent effects on ICM, trophoblast components of embryo outgrowths are also affected. In addition to decreased cell numbers[33], trophoblast cells have abnormally large peripheral hyaloplasmic regions which are most frequently observed during the early, rapid phase of outgrowth (120–167 h p.c.). Peripheral hyaloplasmic regions of control outgrowths exhibit ruffling membrane activity which is rapidly inhibited by CB[30]. The presence of abnormally large hyaloplasmic regions following CB treatments suggests that CB may interact with microfilaments in treated cells to alter subsequent organization and regulation of hyaloplasmic regions. Thus, neither ICM nor trophoblast cells demonstrate a clear differential susceptibility to CB under present experimental conditions.

SUMMARY

As with any probe, interpretations based on perturbations caused by that probe must be carefully drawn. At present, precise effects of CB on cellular components and cellular function are not comprehensively known. Also, it is helpful to know specific cellular conditions at the time of probe initiation. For example, CB at 10.0 μg/ml induces different states of polyploidy in two-cell embryos depending upon the precise stage of the cell cycle[28,40]. Furthermore, identical CB treatments when used to suppress second, third, or fourth cleavage divisions may cause profound differences in CB reversibility and long term developmental success.

Generally, our findings are both predictable and compatible with previously described effects of CB on cellular function. However, we did find a high degree of recovery (as judged by embryo cavitation) following CB treatments even in the 24 h treatment group. Upon characterizing postblastocyst developmental performance of these long term treated embryos *in vitro*, abnormal latent effects on ICM and trophoblast components were observed. Moreover the extent of these latent effects was related to the duration (6, 12, 18, 24 h) of 4.0 μg/ml CB treatment. This duration–response effect may be due to the degree of ploidy, lack of the requisite number of ICM cells, timing of CB treatment (eight-cell embryos from 68–92 h p.c.), and/or other factors; the hypotheses elaborated above are all testable.

Cytochalasin B is a useful tool to probe a wide variety of cellular and tissue phenomena, especially but not limited to those related to morphogenesis. In order to investigate early mammalian embryogenesis, a number of interesting applications ranging from the effects of CB on gametes[26,29,42] and fertilization[26] to the capability of polyploid embryos to survive both *in vitro*[26,32,33,50] and *in utero*[27,28] have been developed. The present report discusses effects of CB on the morula-to-blastocyst transformation, expansion and function of trophoblast outgrowths, and CB-induced latent effects on postimplantation growth and differentiation following treatments at preimplantation (eight-cell) stages.

ACKNOWLEDGEMENTS

The author gratefully acknowledges the long-term assistance and expertise provided by Dr John R. Baker, Dr John P. Trinkaus, and Dr Dorothea

Bennett. Our research has been supported financially by the South Dakota State University Agricultural Experiment Station (SD-737), National Institutes of Health (HD-06918), and the Lalor Foundation.

Figures 5.1, 5.2, and 5.3 plus Tables 5.1, 5.2, and 5.3 have previously been published in *Experimental Cell Research*. The complete citation is listed as reference number 30 in the reference list. The author wishes to thank the editors of *Experimental Cell Research* and publishers of Academic Press, Incorporated, for permission to reproduce these illustrations.

References

1. Carter, S. B. (1967). Effects of cytochalasins on mammalian cells. *Nature*, 213, 261
2. Aldridge, D. C., Armstrong, J. J., Speake, R. N. and Turner, W. B. (1967). The structure of cytochalasins A and B. *J. Chem. Soc. (C)*, 17, 1667
3. Schroeder, T. E. (1969). The role of 'contractile ring' filaments in dividing *Arbacia* eggs. *Biol. Bull.*, 137, 413
4. Schroeder, T. E. (1970). The contractile ring. I. Fine structure of dividing mammalian (Hela) cells and the effects of cytochalasin B. *Z. Zellforsch.*, 109, 431
5. Burnside, B. (1973). Microtubules and microfilaments in amphibian neurulation. *Am. Zool.*, 13, 989
6. Karfunkel, P. (1972). The activity of microtubules and microfilaments in neurulation in the chick. *J. Exp. Zool.*, 181, 289
7. Schroeder, T. E. (1973). Cell constriction: Contractile role of microfilaments in division and development. *Am. Zool.*, 13, 949
8. Luchtel, D., Bluemink, J. G. and DeLaat, S. W. (1976). The effect of injected cytochalasin B on filament organization in the cleaving egg of *Xenopus laevis*. *J. Ultrastruct. Res.*, 54, 406
9. Schroeder, T. E. (1972). The contractile ring. II. Determining its brief existence, volumetric changes, and vital role in cleaving *Arbacia* eggs. *J. Cell Biol.*, 53, 419
10. Selman, G. G., Jacob, J. and Perry, M. M. (1976). The permeability to cytochalasin B of the new unpigmented surface in the first cleavage furrow of the newt's egg. *J. Embryol. Exp. Morphol.*, 36, 321
11. Spooner, B. S. (1973). Microfilaments, cell shape changes, and morphogenesis of salivary epithelium. *Am. Zool.*, 13, 1007
12. Spooner, B. S. and Wessells, N. K. (1972). An analysis of salivary gland morphogenesis: role of cytoplasmic microfilaments and microtubules. *Dev. Biol.*, 27, 38
13. Mousa, G. Y. and Trevithick, J. R. (1977). Differentiation of rat lens epithelial cells in tissue culture. II. Effects of cytochalasins B and D on actin organization and differentiation. *Dev. Biol.*, 60, 14
14. Spooner, B. S. (1974). Morphogenesis of vertebrate organs. In: J. Lash and J. R. Whittaker (eds.). *Concepts of Development*, pp. 213–240. (Stamford: Sinauer Associates, Inc.)
15. Wessells, N. K., Spooner, B. S., Ash, J. F., Bradley, M. O., Luduena, M. A., Taylor, E. L., Wrenn, J. T. and Yamada, K. M. (1971). Microfilaments in cellular and developmental processes. *Science*, 171, 135
16. Yamada, K. M., Spooner, B. S. and Wessells, N. K. (1971). Ultrastructure and function of growth cones and axons of cultured nerve cells. *J. Cell Biol.*, 49, 614
17. Weiss, L. (1972). Studies on cellular adhesion in tissue culture. XII. Some effects of cytochalasins and colchicine. *Exp. Cell Res.*, 74, 21
18. Schaeffer, H. E., Schaeffer, B. E. and Brick, I. (1973). Effects of cytochalasin B on adhesion and electrophoretic mobility of amphibian gastrula cells. *Dev. Biol.*, 34, 163
19. Davis, A. T., Estensen, R. D. and Quie, P. B. (1971). Cytochalasin B. III. Inhibition of human polymorphonuclear leucocyte phagocytosis. *Proc. Soc. Exp. Biol. Med.*, 137, 161
20. Appleton, J. C. and Kemp, R. B. (1974). Effects of cytochalasins on the initial aggregation *in vitro* of embryonic chick cells. *J. Cell Sci.*, 14, 187
21. Schofield, J. G. (1971). Cytochalasin B and release of growth hormone. *Nature New Biol.*, 234, 215
22. Porter, R. and Fitzsimons, D. W. (eds.). (1973). *Locomotion of Tissue Cells*. Ciba Foundation Symp. 14 (new series). (Amsterdam: Elsevier)

23. Balakier, H. and Tarkowski, A. K. (1976). Diploid parthenogenetic mouse embryos produced by heat-shock and cytochalasin B. *J. Embryol. Exp. Morphol.*, **35**, 25
24. Tarkowski, A. K. (1975). Induced parthenogenesis in the mouse. In: C. L. Markert and J. Papaconstantinou (eds.). *The Developmental Biology of Reproduction*, pp. 107–129. (New York: Academic Press)
25. Harper, M. J. K. and Chang, M. C. (1971). Some aspects of the biology of mammalian eggs and spermatozoa. *Adv. Reprod. Physiol.*, **5**, 167
26. Niemierko, A. and Komar, A. (1976). Cytochalasin B-induced triploidy in mouse oocytes fertilized *in vitro*. *J. Reprod. Fertil.*, **48**, 279
27. Snow, M. H. L. (1975). Embryonic development of tetraploid mice during the second half of gestation. *J. Embryol. Exp. Morphol.*, **34**, 707
28. Tarkowski, A. K., Witkowska, A. and Opas, J. (1977). Development of cytochalasin B-induced tetraploid and diploid/tetraploid mosaic mouse embryos. *J. Embryol. Exp. Morphol.*, **41**, 47
29. Wassarman, P. M., Ukena, T. E., Josefowicz, W. J., Letourneau, G. E. and Karnovsky, M. J. (1977). Cytochalasin B-induced pseudocleavage of mouse oocytes *in vitro*. *J. Cell Sci.*, **26**, 323
30. Granholm, N. H. and Brenner, G. M. (1976). Effects of cytochalasin B (CB) on the morula-to-blastocyst transformation and trophoblast outgrowth in the early mouse embryo. *Exp. Cell Res.*, **101**, 143
31. Granholm, N. H. and Draayer, H. A. (1976). Ultrastructural effects of 4.0 μg/ml cytochalasin B on postblastocyst development in mice. *Proc. S. D. Acad. Sci.*, **55**, 85
32. Granholm, N. H. and Brenner, G. M. (1976). Effects of 4.0 μg/ml cytochalasin B on early mouse embryogenesis. *Proc. S. D. Acad. Sci.*, **55**, 106
33. Granholm, N. H., Brenner, G. M. and Rector, J. T. (1979). Latent effects on *in vitro* development following cytochalasin B treatment of 8-cell mouse embryos. *J. Embryol. Exp. Morphol.* **51**, 97
34. Rector, J. T. (1978). Cell surface changes during early mouse development. *Master of Science Thesis* (Brookings: South Dakota State University)
35. Golbus, M., Bachman, R., Wiltse, S. and Hall, B. D. (1976). Tetraploidy in a liveborn infant. *J. Med. Genet.*, **13**, 329
36. McLaren, A. (1976). Genetics of the early mouse embryo. *Annu. Rev. Genet.*, **10**, 361
37. Astaurov, B. L. (1969). Experimental polyploidy in animals. *Annu. Rev. Genet.*, **3**, 99
38. Beatty, R. A. (1957). *Parthenogenesis and Polyploidy in Mammalian Development*. (London: Cambridge University Press)
39. Kaufman, M. H. (1975). The experimental induction of parthenogenesis in the mouse. In: M. Balls and A. E. Wild (eds.). *The Early Development of Mammals*, pp. 25–44. (London: Cambridge University Press)
40. Snow, M. H. L. (1973). Tetraploid mouse embryos produced by ctyochalasin B during cleavage. *Nature*, **244**, 513
41. Harper, M. J. K. and McGaughey, R. W. (1970). Effect of cytochalasin B on development of rabbit eggs. *Anat. Rec.*, **166**, 315
42. Moskalewski, S., Wojciech, S., Gabara, B. and Koprowski, H. (1971). Crystalloid formation in unfertilized mouse ova under influence of cytochalasin B. *J. Exp. Zool.*, **180**, 1
43. Singer, S. J. and Nicolson, G. L. (1972). The fluid mosaic model of the structure of cell membranes. *Science*, **175**, 720
44. DePetris, S. and Raff, M. C. (1973). Normal distribution, patching and capping of lymphocyte surface immunoglobulins studied by electron microscopy. *Nature New Biol.*, **241**, 257
45. Edelman, G. M. (1976). Surface modulation in cell recognition and cell growth. *Science*, **192**, 218
46. Johnson, M. H., Eager, D., Muggleton-Harris, A. and Grave, H. M. (1975). Mosiacism in organization of concanavalin A receptors on the surface membrane of the mouse egg. *Nature*, **257**, 321
47. Jenkinson, E. J. and Billington, W. D. (1977). Cell surface properties of early mammalian embryos. In: M. I. Sherman (ed.) pp. 235–266. (Cambridge: MIT Press)
48. Rector, J. T. and Granholm, N. H. (1978). Differential concanavalin A-induced aggluti-

nation of eight-cell preimplantation mouse embryos before and after compaction. *J. Exp. Zool.*, 203, 497

49. Rowinski, J., Solter, D. and Koprowski, H. (1976). Change in concanavalin A-induced agglutinability during preimplantation mouse development. *Exp. Cell Res.*, 100, 404
50. Sundqvist, K. G. and Ehrnst, A. (1976). Cytoskeletal control of surface membrane mobility. *Nature*, 264, 226
51. Biggers, J. D., Whitten, W. K. and Whittingham, D. G. (1971). The culture of mouse embryos *in vitro*. In: J. C. Daniel, Jr. (ed.). *Methods in Mammalian Embryology*, pp. 86–116. (San Francisco: W. H. Freeman and Co.)
52. Rafferty, K. A., Jr. (1970). *Methods in Experimental Embryology of the Mouse*. (Baltimore: The Johns Hopkins Press)
53. Hally, D. (1964). A counting method for measuring the volumes of tissue components in microscopical sections. *Q. J. Microsc. Sci.*, 105, 503
54. Biggers, J. D. and Brinster, R. L. (1965). Biometrical problems in the study of early mammalian embryos *in vitro*. *J. Exp. Zool.*, 158, 39
55. Thomson, J. L. and Biggers, J. D. (1965). Effect of inhibitors of protein synthesis on the development of preimplantation mouse embryos. *Exp. Cell Res.*, 41, 411
56. Lewis, W. A. and Wright, E. S. (1935). On the early development of the mouse egg. *Contrib. Embryol. Carnegie Inst. No.*, 148, 115
57. Mulnard, J. G. (1967). Analyse micro-cinematographique du développement d'oeuf de Souris du stade II au blastocyste. *Arch. Biol. (Liége)*, 78, 107
58. Ducibella, T. and Anderson, E. (1975). Cell shape and membrane changes in the eight-cell mouse embryo; prerequisites for morphogenesis of the blastocyst. *Dev. Biol.*, 47, 45
59. Wessells, N. K., Spooner, B. S. and Luduena, M. W. (1973). Surface movements, microfilaments, and cell locomotion. In: R. Porter and D. W. Fitzsimons (eds.). *Locomotion of Tissue Cells*, pp. 53–77. Ciba Foundation Symposium 14 (new series). (Amsterdam: Elsevier)
60. Estensen, R. D., Rosenberg, M. and Sheridan, J. D. (1971). Microfilaments and 'contractile' processes. *Science*, 175, 774
61. Sanger, J. W. and Holtzer, H. (1972). Cytochalasin B: effects on cell morphology, cell adhesion, and mucopolysaccharide syntheses. *Proc. Natl. Acad. Sci. (USA)*, 69, 253
62. Lin, S. and Spudich, J. A. (1974). Biochemical studies on the mode of action of cytochalasin B. *J. Biol. Chem.*, 249, 5778
63. Yamada, K. M. and Wessells, N. K. (1973). Cytochalasin B: effects on membrane ruffling, growth cone and microspike activity, and microfilament structure not due to altered glucose transport. *Dev. Biol.*, 31, 413
64. Mayhew, E., Poste, G., Cowden, M., Tolson, N. and Maslow, D. (1974). Cellular binding of ^{3}H-cytochalasin B. *J. Cell Physiol.*, 84, 373
65. Miranda, A. F., Godman, G. C., Deitch, A. D. and Tanenbaum, S. W. (1974). Action of cytochalasin D on cells of established lines. I. Early events. *J. Cell Biol.*, 61, 481
66. Miranda, A. F., Godman, G. C. and Tanenbaum, S. W. (1974). Action of cytochalasin D on cells of established lines. II. Cortex and microfilaments. *J. Cell Biol.*, 62, 406
67. Smith, R. and McLaren, A. (1977). Factors affecting the time of formation of the mouse blastocoel. *J. Embryol. Exp. Morphol.*, 41, 79
68. Perry, M. M. and Snow, M. H. L. (1974). The blebbing response of 2–4-cell mouse embryos to cytochalasin B. *Dev. Biol.*, 45, 372
69. Tarkowski, A. K. and Wroblewska, J. (1967). Development of blastomeres of mouse eggs isolated at the 4- and 8-cell stage. *J. Embryol. Exp. Morphol.*, 18, 155
70. Bennett, D. (1975). The T-locus of the mouse. *Cell*, 6, 441
71. Sherman, M. I. and Wudl, L. R. (1977). T-complex mutations and their effects. In: M. I. Sherman (ed.). *Concepts in Mammalian Embryogenesis*, pp. 136–234. (Cambridge: The MIT Press)
72. Granholm, N. H. and Johnson, P. M. (1978). Identification of eight-cell t^{w32} homozygous lethal embryos by means of aberrant compaction. *J. Exp. Zool.*, 203, 81
73. Spiegelman, M. (1976). Electron microscopy of cell associations in T-locus mutants. In: K. Elliott and M. O'Connor (eds.). *Embryogenesis in Mammals*. Ciba Foundation Symposium 40 (new series), pp. 199–220. (Amsterdam: Elsevier)
74. Spiegelman, M. and Bennett, D. (1974). Fine structural study of cell migration in the early

mesoderm of normal and mutant mouse embryos (T-locus: t^9/t^9). *J. Embryol. Exp. Morphol.*, **32**, 723
75. Tung, K. H. (1978). Observations on attachment and spreading of zonaless t^{w32}/t^{w32} and control preimplantation mouse embryos. Master of Science Thesis. (Brookings: South Dakota State University)
76. Sherman, M. I. (1974). *In vivo* and *in vitro* differentiation during early mammalian embryogenesis. *Front. Rad. Therapy*, **9**, 122
77. Sherman, M. I. and Salomon, D. S. (1975). The relationship between the early mouse embryo and its environment. In: C. L. Markert and J. Papaconstantinou (eds.). *The Developmental Biology of Reproduction*, pp. 277–309. (New York: Academic Press)
78. Spindle, A. I. and Pedersen, R. A. (1973). Hatching, attachment, and outgrowth of mouse blastocysts *in vitro*. *J. Exp. Zool.*, **186**, 305
79. Ansell, J. D. and Snow, M. H. L. (1975). The development of trophoblast *in vitro* from blastocysts containing varying amounts of inner cell mass. *J. Embryol. Exp. Morphol.*, **33**, 177
80. Snow, M. H. L. (1976). The immediate postimplantation development of tetraploid mouse blastocysts. *J. Embryol. Exp. Morphol.*, **35**, 81
81. Epstein, C. J. (1975). Gene expression and macromolecular synthesis during preimplantation embryonic development. *Biol. Reprod.*, **12**, 82
82. Van Blerkom, J. and Manes, C. (1977). The molecular biology of the preimplantation embryo. In: M. I. Sherman (ed.). *Concepts of Mammalian Embryogenesis*, pp. 37–94. (Cambridge: MIT Press)
83. Pedersen, R. A. and Spindle, A. I. (1976). Genetic effects on mammalian development during and after implantation. In: K. Elliott and M. O'Connor (eds.). *Embryogenesis in Mammals*. Ciba Foundation Symposium 40 (new series), pp. 133–149 (Amsterdam: Elsevier)
84. Brock, K. T. (1978). Effects of A^y on preimplantation embryo lethalities (A^y/A^y) and on abnormal mating behavior (A^y/a) in strain C57BL/6J mice. *Master of Science Thesis* Brookings: South Dakota State University)
85. Johnson, L. L. and Granholm, N. H. (1979). *In vitro* analysis of pre- and early postimplantation development of lethal yellow (A^y/A^y) mouse embryos. *J. Exp. Zool.*, **204**, 38
86. Glass, R. H., Spindle, A. I. and Pedersen, R. A. (1976). Differential inhibition of trophoblast outgrowth and inner cell mass growth by actinomycin D in cultured mouse embryos. *J. Reprod. Fertil.*, **48**, 443
87. Rowinski, J., Solter, D. and Koprowski, H. (1975). Mouse embryo development *in vitro*: Effects of inhibitors of RNA and protein synthesis on blastocyst and post-blastocyst embryos. *J. Exp. Zool.*, **192**, 133
88. Sherman, M. I. and Atienza, S. B. (1975). Effects of bromodeoxyuridine, cytosine arabinoside and colcemid upon *in vitro* development of mouse blastocysts. *J. Embryol. Exp. Morphol.*, **34**, 467

6
Pathogenesis of cyclophosphamide-induced fetal anomalies

D. WENDLER

INTRODUCTION

During the past 20 years, cyclophosphamide has been used increasingly as a model teratogen for studying the response of embryonic tissues or as a reference material for determining the sensitivity of different strains of animals. Numerous studies have been carried out on the teratogenic effects of cyclophosphamide in different species of animals: chicken[1-4], mouse[5-9], rat[10-21], and rabbit[1,22-25]. These authors investigated the effect of cyclophosphamide, particularly on different days of embryonic development during which the conceptuses are particularly susceptible and the most impressive anomalies are induced. In contrast, information on the changes induced by cyclophosphamide during late pregnancy have received relatively little attention[9,15,26]. For man, too, there have been reports that cyclophosphamide treatment interferes with fetogenesis (for review, see refs. 27 and 28).

Even during the first days of life cyclophosphamide or the structurally analogous isofamide induce disturbances in the differentiation of definite organs or growth retardation, including anomalies of the teeth[29,30], anomalies in the structure of the internal ear[29,31], developmental defects of the photoreceptors of the retina[32], disturbances of the growth of the skull[29,33], anomalies of the forelimbs[33], and general disturbances in maturity[34].

Histological methods provided additional information with respect to the formal teratogenesis of definite malformation syndrome. Such studies were carried out in single organs after application of cyclophosphamide during the embryonal and fetal phases of development[17,20,26,35], followed by histochemical or autoradiographic methods, respectively[36,37]. Similar investigations are to be found in literature for N-mustard[38], N-methyl-N-nitrosourea[17], N-ethyl-N-nitrosourea[26], different hydrazides, carbonic acid amides, and hydroxamic acids[35,39], methotrexate, acetazolamide, hydroxyurea[40], and 5-azacytidine[41]. Our investigation represents a contribution to the formal genesis of cyclophosphamide-induced fetal anomalies in the rat based on histological analysis of the developing malformation syndrome.

MATERIALS AND METHODS

Wistar albino rats, kept under conventional conditions, were used for the present investigation. The mature females (200–250 g) were mated overnight in the proportion of two males to six females. Evidence of sperm-positive vaginal smears was taken as the first day post coitum (p.c.). A total of 658 conceptuses obtained from 47 animals at intervals of 24 h from the 15th to the 21st gestational day, served as controls. The test groups consisted of a total of 53 animals, which received intraperitoneally on the 15th and 16th day p.c. a previously determined optimal teratogenic dose of 20 mg cyclophosphamide (CPA)/kg body weight. At 24 h intervals, a number of mothers were killed with an overdose of chloroform and the fetuses were dissected out. Thus, conceptuses from the 16th to the 21st day were available for histological studies. After careful external inspection by means of a stereomicroscope and measurement of the length, the fetuses were weighed, and then fixed in a mixture of alcohol (96%), formol (30%) and acetic acid (in the proportion of 12:6:1). Half of the fetuses recovered on the 21st day p.c. were fixed in 96% alcohol and eviscerated for subsequent skeletal staining (alizarin S). Histological examination was carried out on paraffin serial sections which were stained preferably with haematoxylin and eosin. In order to demonstrate clearly necrotic cells, we used Kernechtred combined with alcian blue (pH 3), and the PAS reaction was used to demonstrate glycogen and neutral mucopolysaccharides. The objectives of these studies were not to report general parameters of reproduction or exact data on the frequency of individual malformations developing (implantation and resorption rates, number of living and dead fetuses, retardation and rates of malformations, sex ratio and fetal weights). For this reason a statistical evaluation of the data is not included.

RESULTS

Subdivision of prenatal developmental processes can be attempted according to different criteria: morphogenetic stages, type of nutrition, and metabolic or immunological criteria. The subdivision that we have used takes into consideration both the structural criteria of morphogenesis and differentiation and the sensitivity and reaction of the conceptuses under experimental conditions. In the rat, the fetal period of development can be further subdivided into *early fetogenesis* (15th and 16th day p.c.) and *late fetogenesis* (17th day p.c. until birth). The early fetal period, during which single conceptuses, by the variability of intrauterine development can react 'still' as an embryo or 'already' as a fetus is the most interesting. Before discussing the malformation syndromes that were induced in early fetogenesis and their formal pathogenesis, the tissues that were likely to be sensitive on the basis of morphological criteria will be briefly reviewed.

Degree of development in rat fetuses from the 15th to the 16th day p.c.

Early fetogenesis is marked by extensive differentiation of all organ systems developing during early (7th to 11th day p.c.) and late (12th to 14th day p.c.) embryogenesis, by the formation of some essential new fetal structures, and by the regression of some embryonal organs.

Body shape
The typical embryonal curvature gives way to erection of the trunk. At the end of the 16th day p.c., 96% of the fetuses have undergone an extension.

Nervous system
Further development and differentiation of the cerebrum. Formation of layers in the cortex, nuclear isolations in the diencephalon and the rhombic fossa, anlage of epiphysis and paraphysis. Formation of the medulla of the adrenal gland from neuroblasts of the neural crest.

Eye
The embryonal choroidal fissure is closed. The lid folds which in the beginning have a circular course around the bulbus grow towards each other from the 16th day p.c. Epithelial fusion of the eyelids, nevertheless, occurs on the 19th day p.c. only.

Ear
Formation of the auricle by fusion of the mesenchymal auricular appendages. Chondrogenesis begins in the auricle and in the external meatus. The transitory closure of the external ear by outgrowth of the auricle with the skin occurs from the beginning of the 18th day p.c.

Nasal and oral cavities
On the 15th day p.c. the nasal septum lengthens and fuses with the two palatal processes of the maxillary process under formation of the final palate at the end of the 16th day. The external nostrils are closed in all fetuses by an epithelial plug from the beginning of the 15th day p.c.

Mandible
Up to the 16th day p.c., the lower jaw remains considerably shorter than the maxilla and the tongue protrudes out of the oral cavity. On the 17th day p.c., 20% of the fetuses still show these characteristics, while on the 18th day p.c. the lower and upper jaws are equal and the tongue has shifted back into the oral cavity.

Urinary tract
Beginning regression of the parts of the mesonephros which do not grow into epididymis in male fetuses.

Genital tract
Formation of the Müllerian duct. Regression of the cloacal membrane.

Heart and circulation system
Complete separation into left and right heart.

Chondrogenesis
Formation of the skeletal anlage out of the corresponding mesenchymal blastemas.

Extremities
At the *forelimbs*, a proximal and a distal part are already visible on the beginning of the 13th day. On the 15th day p.c., the finger rays in the handplate appear which, on the 16th day, are connected by 'interdigital webs'. The latter regress on the beginning of the 17th day p.c. On the 16th day p.c., introversion of the handplate begins so that the thumb is turned in the medial direction. The *hindlimb* is divided into a proximal and a distal segment on the 15th day p.c., and the footplates are visible. On the 16th day p.c., the toe rays, which are already visible externally, are formed. They are all connected by 'interdigital webs' on the 17th day p.c.

Physiological omphalocele
While the physiological omphalocoele is evident in all fetuses on the 15th day p.c., this characteristic is already absent in 86% of all fetuses (on the 18th day p.c.).

Development of weight
Table 6.1 shows the weight increase in our strain of rats during fetogenesis.

Table 6.1 Weight development of the rat in fetogenesis

Day p.c.	$\bar{x} \pm s_{\bar{x}}$
15	0.188 ± 0.002
16	0.311 ± 0.002
17	0.501 ± 0.006
18	0.907 ± 0.012
19	1.423 ± 0.019
20	2.209 ± 0.014
21	3.243 ± 0.029
22	4.813 ± 0.049
23	5.636 ± 0.061

$\bar{x} \pm s_{\bar{x}}$ = mean and standard error in grammes

Malformation syndromes on the 21st day p.c. after application of cyclophosphamide on the 15th day p.c.

After application of CPA on the 15th day p.c., the malformation rate was 100%. All fetuses were externally malformed, and had skeletal anomalies; 60% revealed malformations of the internal organs.

The severe toxic effects of the drug were evident from the oedema and often large fluid-filled skin blebs, often containing blood (Figure 6.1). Haemorrhages

CYCLOPHOSPHAMIDE TERATOGENICITY

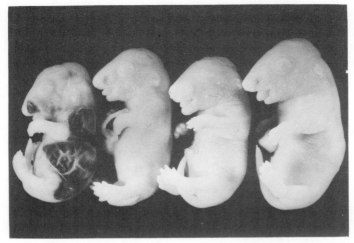

Figure 6.1 All rat fetuses show exophthalmia, ablepharia, micrognathia, and oedema on day 21. In the left fetus there are blood-filled skin blebs and haemorrhage of the touch hair buds. CPA given on day 15 p.c. × 1.71

were present on the entire body surface and within the region of serous membranes. A typical localization of haemorrhages were the anlages of touch-hair buds and the tissues surrounding both kidneys. Impressive malformations were found in the region of the head: 100% of the fetuses showed an occipital haemorrhagic encephalocele on the 21st day p.c. This was located particularly in the region of the parietal bones, which in most cases were entirely absent (Figure 6.2). The defect was covered by the extremely thin transparent skin. Whereas ossification of the frontal bone was mostly normal, the anlages above all the occipital bone showed considerable retardations: the supraoccipital and

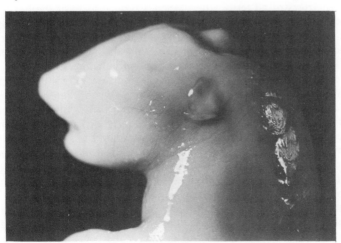

Figure 6.2 Encephalocoele in the region of the midbrain. Lack of the osseous neurocranium. Micrognathia. Rat fetus on day 21. CPA given on day 15 p.c. × 4.77

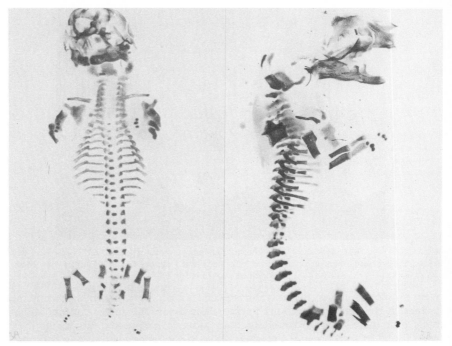

Figure 6.3 Skeletal preparations of rat fetuses on day 21. CPA given on day 15 p.c. × 3.4: a, Cleft palate, short ribs, knobby swellings of the costal angle, short fibula; b, Lack of the neurocranium. Disturbances of the thoraco-lumbar connection

interparietal bones were absent (Figure 6.3B) or both anlages were not yet fused together.

Three-quarters of all fetuses developed a typical form of the head, marked by trunk-shaped maxillae and convex bridges of the nose (Figure 6.1); in the controls, the frontal and nasal bones were on the same level but this part projected convexly in animals treated with CPA. This shape of the skull, named 'Rammskopf' is based on the shift of proportions of different sections of the cerebrum (hypoplasia of the pallium and normal configuration of the olfactory bulb). Nearly 80% of the fetuses showed moderate to severe micrognathia as a result of which the tongue protruded much out of the opened oral cavity (Figure 6.1). In 20% of the fetuses, bilateral cleft palate developed with not yet erected palatal shelves. After examination of the skeletal preparations, it was found, however, that in 40% of all fetuses in which the palate seemed to be closed (razor blade technique), the centres of ossification were still far from the median line; whether or not this defect was due to retardation, a true malformation can develop secondarily by rupture of the thin membrane.

Developmental arrest of the eyelids and flattening of the orbit caused exophthalmia to appear in the fetuses examined on the 21st day p.c. (Figure 6.1). The bulbs protruded much out of the orbits. Turbidities and hydropic swelling of the cornea suggested degenerative changes. The eyelids are either completely absent (ablepharia) or are circularly arranged around the bulb. Lid fusion, normally appearing on the 19th day p.c. was absent in 75% of all fetuses (7% unilateral, 68% bilateral). The structure of the retina seemed to be normal. We saw in only one case a folded retina and agenesis of the lens. The auricles in most fetuses were located too low (Figure 6.1), perhaps resulting from the oedemas of the scalp. The anlages of the membranous labyrinth were microscopically normal.

Disturbances in the closure of the neural tube rarely appeared: in two fetuses (approximately 1%), the fourth ventricle was connected with the surface of the head. In this case the external skin continuously extended to the neural tissue. Because the closure of the caudal neuropore in the rat already takes place on the 12th day p.c., the defect which occurred can only be explained as secondary rupture.

Disturbances in the formation of the vertebral canal appeared in the form of spina bifida occulta, particularly in the thoracic and lumbar regions. Because fusion of the two anlages of spinal processes was absent as a result of general skeletal retardation in nearly half the fetuses, the spinal cord was directly connected to the skin in certain regions. Nevertheless, the existence of a hernia-like protrusion of a posterior horn of the grey matter of the spinal cord in some cases did not support their classification as a retardation. Wrist-drop bendings of the forepaws appeared sometimes at the limbs without evidence of defects in the bone anlages (Figure 6.1). Twenty-five percent of all fetuses showed anomalies in the rotation of the posterior limbs with abnormal extension and sometimes supinated hindfeet (clubfoot). Of the bones of the leg in all fetuses, only short fibulae were evident, occupying the distal third and ending at a point proximally (Figure 6.3a). Three percent of the fetuses had anomalies in the number of toes. These oligodactylias were presented by fetuses which on the day of treatment (15th day p.c.) were still at the developmental stage of the 14th day because of natural variability. At this stage, the distal section of the posterior extremities is still sensitive to CPA. Thus the anomalies in the number of fingers or toes do not belong to the complex of fetopathia.

All developmental anomalies present on the vertebral column must be classified as retardations, lacking ossification centres of the lower thoracic and the upper lumbar vertebral column with the existence of cartilaginous anlages (Figure 6.3b), termination of the bone nuclei in the sacral region, pale and porous centres of ossification. At the vertebral column of the tail, the apparent maldevelopment of the cartilaginous anlages caused, nevertheless, in half the fetuses kinky tails of different degrees. On the thoracic skeleton, retardations by shortening and incomplete ossification of the ribs, as well as frequent absence of the 13th pair of ribs, became evident. A knotty swelling at the costal angle (Figure 6.3a) or an S-shaped distortion of the ribs was often found.

We seldom observed malformations of the internal organs following CPA treatment, with the exception of the central nervous system and the eye: 5% hydronephroses, 0.5% dystopias of kidneys, 3% gastroschisis. Treated fetuses have an average body weight of $\bar{x} = 2.12$ g on the 21st day p.c. This decrease in weight, differing from the control value ($\bar{x} = 3.24$ g) by 35%, lies within the double variation of the weights at the 20th day p.c. (Table 6.1). Thus, the results are that certain features of the fetuses, which develop between the 20th and 21st day p.c. only, cannot be considered as malformations (shift of head–body proportion in favour of the body, presence of an angle between the frontal and the nasal bones, lack of ossification centres, especially of the caudal vertebral column, the metatarsal and the sternal bones, no anlage or fusion of bone nuclei in the osseous skull).

The size of the fetuses decreased with their weight whereas control fetuses had a mean length of $\bar{x} = 3.66 \pm 0.03$ cm on the 21st day p.c., it was only 72% ($\bar{x} = 2.63 \pm 0.05$) of the control value after treatment on the 15th day p.c.

Malformation syndromes on the 21st day p.c. after application of cyclophosphamide on the 16th day p.c.

After application of CPA on the 16th day p.c. the malformation syndromes observed were different from those seen after CPA application on the 15th day p.c. The occipital encephalocoeles were no longer part of the typical malformation syndrome of the 16th day. Very isolated herniations of the brain can be explained as a normal retardation of 24 h of the corresponding fetuses. The frequent 'Rammsköpfe' with elevated bridges of the nose were absent after application of CPA on the 16th day p.c. Hypoplasias of the jaw skeleton with medium brachygnathias and shortened, trunk-like upper jaws developed which often resulted in a severe deformity at the root of the nose. The tongue protruded out of the oral cavity. The eyes were closed following fusion of the epithelial lid but bulged through the extremely thin skin above the surface. Skeletal staining revealed very flat orbital bones similar to application of CPA on the 15th day p.c. In all fetuses the head was erect. Bending of the vertebral column commenced at the cervico-thoracic outlet. In individual cases, anomalies of the limbs were observed: fairly extended fore- and hindlimbs or clubfeet. The number of fingers and toes was always normal.

There were only few visceral malformations, including hydronephrosis and renal dystrophia. The CPA treatment caused severe retardation of the skeletal system. The skull was most affected; whereas all bone centres of the facial skull and of the jaw developed, the osseous neurocranium was entirely absent or existed only partly on the 21st day p.c. (parietal, supraoccipital, interparietal bones). When the frontal squama was also missing, the whole calvarium was absent and the encephalon bordered on the intact skin with its broad surface. In the region of the vertebral column, especially in the thorax, single bony anlages of vertebral corpora were absent (when the corresponding cartilaginous elements were present). The same was true for the lateral bony centres which often appeared very porous and shadowy in the lumbar region. Compared to the application of CPA on the 15th day p.c., ossification of the ribs increased considerably even if calcification in single regions only partly existed. In most cases, S-shaped bends were noticeable at the costal angle, but they were less in contrast to those observed after application of CPA on the previous day of development.

The palate was closed on the 21st day p.c. (razor blade technique). However, skeletal staining showed that in all fetuses there was either no evidence of ossification or the ossification centres of both sides were still far distant from the median line. Poor skeletal development was observed in the sternal, clavicular, metatarsal, pelvic, and hyoid bones. Nonspecific toxic appearances in the form of vitreous subcutaneous oedemas and of large fluid-filled skin blebs have increased, compared to the application of CPA on the 15th day p.c.

Formal teratogenesis of the CPA-induced fetopathia after application on the 15th day p.c.

Fetuses were recovered after treatment with CPA on the 15th day p.c. at intervals of 24 h and prepared for histological studies. Microscopic examination of the placentas and of the visceral yolk sac epithelium, essential for nourishment, gave no indication that the damage to the conceptuses resulted secondarily from an impairment of fetal nutritional functions.

24 h after application of CPA
The fetuses corresponded macroscopically to those of the 16th day of development without the signs of beginning anormogenesis. Histologically, degenerative processes were involved which included karyopycnosis, karyorhexis and karyolysis. The degenerating cells became separated from the arranged syncytium and rounded off. Frequently, a pycnotic cell nucleus decayed into fragments before chromatolysis (degeneration granules). These were equally scattered as basophilic nuclear debris over the degenerated tissue. In several affected blastemas, large acinous necrotic conglomerates (necrotic cells) appeared, the pooled cytoplasm of which was fairly acidophilic and PAS-positive. In the case of condensation, the impression of multinuclear giant cells is conveyed. Despite careful examination in most cases no morphologically (vital) cell nucleus could be identified.

Damage done to ectodermal derivatives
The central nervous system was particularly involved. In contrast to the structurally normal pallium of day 16 of development (Figure 6.4a), a strong disaggregation followed application of CPA (Figure 6.4b). The neuroepithelial layer, close to the ventricles appeared to be condensed without unequivocally proven mitoses. The mantle layer was fairly disaggregated as a result of massive cell destruction and oedema in the tissue; a sequence of layers could not be identified. Numerous degeneration granules and acinous necrotic foci occupied this zone. Visible reticular structures corresponded to glial elements and blood capillaries. A clear delineation of the hemisphere from mesenchyme of the skin, as is found in normal fetuses by the condensation of the scalp, did not exist (Figure 6.4a). Few inconstant necrotic foci were arranged in the wall of the olfactory bulb and in the anlage of the hippocampal region. Besides typical degeneration granules, many necrotic conglomerates were also found in the ganglionic colliculus, not yet subdivided (Figure 6.5). In the region of the diencephalon, the necrotic foci occupied a dorsal middle zone of the ventricular wall which is located above the sulcus limitans; the bottom of the diencephalon seemed to be intact (Figure 6.6). Several degenerating areas were also found in the region of the mesencephalon. The disaggregation and destruction of cells was mostly located in the roof of the mesencephalon. In the wall of the fourth ventricle, similar changes were observed.

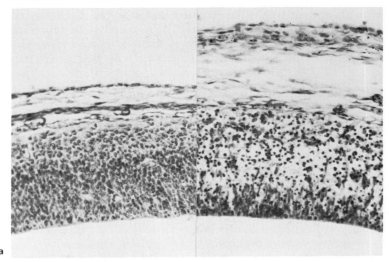

Figure 6.4 Sagittal section of the brain cortex on day 16 p.c. H & E × 155: a, Normal cortical layers with surrounding desmocranium; b, Damaged brain cortex following application of CPA on day 15 p.c. Degeneration granules and necrotic conglomerates in the mantle layer. Lack of desmocranium and oedema

Here also, the dorsal parts were more affected than the bottom of the rhombic fossa. As in all other sections of the brain, the ependymal layer remained unaffected in structure.

The cervical and thoracic regions of the spinal cord were barely injured. At the thoracolumbar level, degeneration foci appeared which in some cases

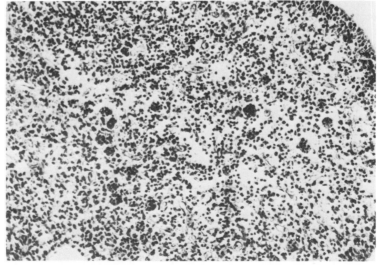

Figure 6.5 Degeneration and eosinophilic conglomerates in the striatal portion of the telencephalon on day 16. CPA given on day 15 p.c. H & E × 160

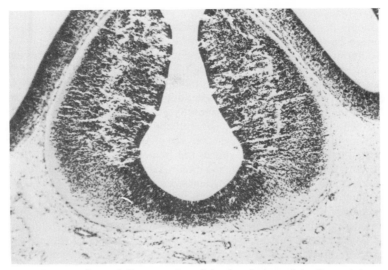

Figure 6.6 Degeneration and disaggregation of the dorsal parts of the diencephalon on day 16. CPA given on day 15 p.c. H & E × 45

extended to the caudal spinal cord with varying intensity: cellular necroses, acinus conglomerates and oedema disaggregate the spinal cord dorsally to the sulcus limitans (Figure 6.7). The area of the basal lamina and the spinal ganglia were not affected. The undifferentiated neural tube in the region of the tip of the tail showed severe regression. Other ectodermal derivatives (epi-

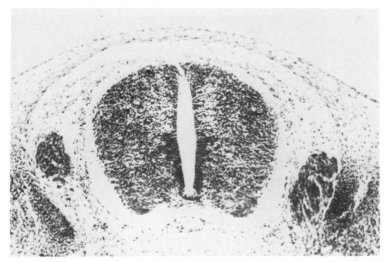

Figure 6.7 Degeneration and disaggregation in the alar lamina of the spinal cord on day 16. The matrix layer is ventrally preserved. Surrounding oedema in the primitive meninx and skin. CPA given on day 15 p.c. H & E × 60

dermis, retina, pigment layer of the eye, crystalline lens and the derivatives of otic placode) developed according to their age.

Damage done to mesodermal derivatives

The proliferation of the mesodermal mesenchyme and of the blastemas derived from it (desmo-, myochondroblastemas) was affected by CPA depending on the location. The head mesenchyme over pallium and mesencephalon was particularly involved. Increased blood supply, massive cell destruction and tissue fluid invasion enlarged the mesenchymal network (Figure 6.4b). Foci of necrotic cells were located in the subepithelial connective tissue around the oral and nasal cavities, in the mesenchyme of the maxillary and mandibular processes, in the nasal septum and in the free ends of the palatal processes. The latter were inhibited in their kinetics, still lying vertically beside the tongue (normally fusion occurs in the course of the 16th day). The mesenchyme, surrounding the touch hair buds was especially affected: the epithelial buds were surrounded by 'layers' of necrotic cells. Furthermore, the auricular anlages and eyelids represented sites of massive cell destruction. In the mesenchyme of the dorsal auricular hillock, outside the anlage of the prospective cartilaginous blastema, and in the upper lid, very severe degeneration was observed. Although the connective tissue of the cornea was oedematous, there was no evidence of cell degeneration.

The mesenchyme of the skin on the trunk was affected in the following manner: sites of degeneration and oedematous swelling, located just under the basal membrane separated the otherwise dense tissue. The dorsal skin and the maxillary process were especially involved. Numerous necroses were found in the condensed mesenchyme of the genital tubercle. Similar to the mesenchyme above the pallium, the primitive covering of the spinal cord was severely swollen (Figure 6.7). The mesenchyme of the tail, too, was particu-

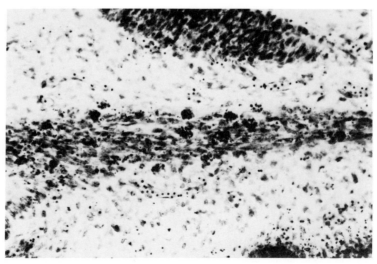

Figure 6.8 Necrotic conglomerates in muscles of the abdominal wall on day 16. CPA given on day 15 p.c. H & E × 160

larly affected by CPA: necroses were histologically evident parachordally and in the condensed parts of the rudimentary somites of the tail.

The degeneration granules and necrotic centres, which were seen on the 16th day p.c. in the anlages of the extremities, did not differ in intensity and localization from those which appear as physiological necroses. However, disaggregations of the tissue (separation of the loosely arranged parts of the condensed future cartilaginous blastemas) were found in the distal segments of the extremities. The following muscle blastemas revealed particularly strong sensitivity when large necrotic sites appeared: ventral cervical muscles, thoracic and abdominal muscles (Figure 6.8), dorsal diaphragm, and the tongue.

48 h after application of CPA

Macroscopic examination of the fetuses on the 17th day p.c. revealed no real evidence of abnormalities. In certain blastemas, degeneration continued, whereas in other regions repair was prominent.

Damage done to ectodermal derivatives

In the central nervous system, degeneration granules and necrotic sites were located near to each other. The damaged parts of the wall of the brain increased in thickness from oedema and haemorrhages in the necrotic sections, as a result of which the lateral ventricles were compressed. Also the choroid plexus became filled with blood; the mantle layer of the isocortex was most severely damaged, but the accumulation of blood and cell debris, observed in all ventricles, also disturbed the structure of the neuroepithelial layer. It could be interpreted as a sign of beginning reparation that within the necrotic layers cavities appeared by removal of cell debris (Figure 6.9). In the cavities, round

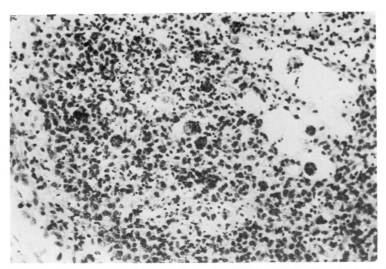

Figure 6.9 Concurrent events of degeneration and reparation on day 17 in the brain. Fetal macrophages are surrounded by small or large cavities. CPA was given on day 15 p.c.
H & E × 160

eosinophilic masses of cytoplasm were sometimes found in which many basophilic particles, sometimes even a structurally normal nucleus, were recognizable. The latter suggested that fetal macrophages were responsible for enlarging the cavities by removal of the cell debris. Similar lesions affected the striatal portion, the dorsal nuclei of the thalamus, and the dorsal roof of the mesencephalon as well as the roof of the fourth ventricle, which could be completely damaged.

Damage done to mesodermal derivatives

Cellular degeneration in the subcutaneous connective tissue was completed. Large labyrinthine spaces were found in the mesenchyme as a result of which the surface epithelium often appeared (light-microscopically) without connective tissue support. Epithelial structures in the region of these beginning skin blebs (glands, hair anlages, Figure 6.10) traversed the hollow spaces.

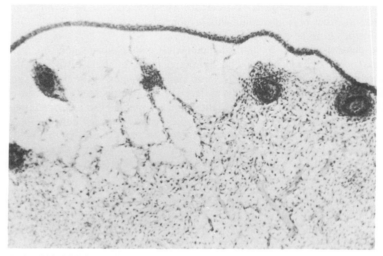

Figure 6.10 Skin bleb in the region of touch hairs on day 17. After degeneration the overlying epidermis seems to be free of mesenchyme. CPA given on day 15 p.c. H & E × 63

Mesenchymal cyst formation was also observed at the free ends of the vertically oriented palatal processes. Swollen mesenchyme covered the brain anlage also, especially in the region of the mesencephalon. Necrotic processes continued in the oedematous eyelids, in the anlage of the auricles, in the mesenchyme of the maxilla and mandible, and in the tail. In some fetuses, the cartilaginous blastema of the future sternal bone was absent.

72 h after application of CPA

On the 18th day p.c. the fetuses treated with CPA reached a degree of development which corresponded to the 17th day p.c. Besides this intrauterine growth retardation, the malformation syndromes typical for the 21st day p.c. were already macroscopically visible. All fetuses showed an oedematous swelling but fewer skin blebs and haemorrhages than later on; brachygnathias, convex bridges of the nose, and marked curvature of the mesencephalon were present in all fetuses. Eyes and ears were opened, and most of the fetuses showed early signs of a curved tail.

The histological appearance was determined by the reparation processes. Necrotic cell debris was found in the ventricles only. Hollow cysts were scattered in the cortex but in other regions of the brain led to considerable loss of brain tissue (Figure 6.11). Macrophages with numerous basophilic granules were still present but gradually disappeared. In several regions of the cerebral cortex, extensive blood extravasates were still present which compressed the lateral ventricles together with disaggregation of tissue. As an indication of regeneration, layers of neuroblasts were found settled above the disaggregated mantle zone of the pallium.

Formation of cavities and haemorrhages were also observed in the striated portion, subdivided by the anlage of the internal capsule. Above the mesencephalon a structurally closed desmocranium was lacking. The cystic transformation, taking place in certain regions of the body, progressed (formation of blebs at the trunk and in the areas of the touch hairs and eye lids, Figure 6.12). Formation of cysts was also found in the non-united palatal processes and in the nasal septum. Of special interest, several fetuses lacked the condensed connective tissue layers bordering the vertebral canal, which eventually led to spina bifida occulta.

DISCUSSION

The adverse effects of environmental chemicals or drugs during late pregnancy are increasing in importance. While the earlier teratological test schedules of the pharmaceutical industry often ignored the fetal period, the findings of experimental teratology have fundamentally altered our opinion with respect to the sensitivity for, or resistance of, fetogenesis to chemical agents. As some of the adverse effects induced during fetal development cannot be determined by morphological techniques, it is now required that the maturation of vital organ systems, especially of the central nervous system, be included in drug research. Only the essential aspects of this area of 'behavioural teratology' are presently formulated and it needs further and exact guidelines (for review see refs. 42 and 43). Studies with cyclophosphamide[9,15,26,44], lathyrogenic compounds[9,44], procarbazine[45], 5-iododeoxyuridine and cytosine arabinoside[46], N-methyl-N-nitrosourea and N-ethyl-N-nitrosourea[26], showed that fetal tissues are not inferior to the embryonic tissues in their ability to produce malformations. Fetopathias were also observed after ligation of the uterine vessels[47], after amniocentesis[48], by irradiation[49], and by viral infections[50].

The pathogenesis of cyclophosphamide-induced fetal abnormalities has been described in detail[26]. These observations are, to a large extent, in agreement with our results. Both authors treated pregnant rats on the 15th or 16th day p.c. with 20 or 40 mg cyclophosphamide/kg body weight, respectively, and investigated histologically the pathogenesis of the resulting malformation syndrome. Compared with other cyclophosphamide-induced damages during the embryonic period[14–17,20,51], different malformation patterns were seen

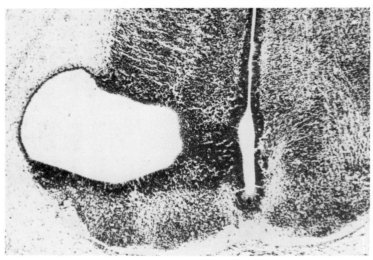

Figure 6.11 Defect of neural substance in the midbrain region after complete removal of necrotic material on day 18. CPA given on day 15 p.c. H & E × 45

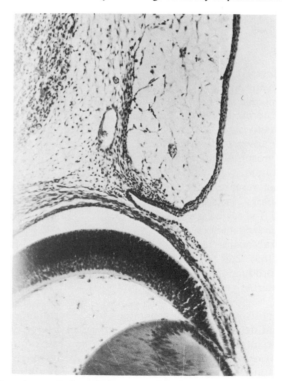

Figure 6.12 Growth retardation of the upper lid after termination of necrotic processes. Oedema of the subcutaneous mesenchyme on day 18. CPA given on day 15 p.c. H & E × 70

which suggested that in fetogenesis single organs no longer react, but that other tissues have become sensitive to the agent. While the adverse effects during embryogenesis were dose-dependent and can involve the entire brain, the described cell necroses seen after application of cyclophosphamide in the early fetal stage of development were restricted to the dorsal aspects of the cerebral hemispheres, the striatal portion and to the sensory alar plates of the brainstem and the spinal cord. In the last third of pregnancy, cell proliferation, neural differentiation, and cell migration are characteristic events in the central nervous system[41]. These events do not begin at the same time in the whole brain, but take place on definite developmental days of fetogenesis in definite areas[41,52]. Thus teratogenesis is dependent on the day of treatment and results in deficiencies of cell number, minute cell abnormalities, and cell dystopia[41]. Inhibition of normal cell migration out of the multilayered neuroepithelial zone plays a special role. The proliferative neuroblasts replicate their DNA outside the neuroepithelial layer and then migrate towards the lumen in order to divide[41]. While some neuroblasts repeat the cycle, others migrate into the mantle layer to differentiate therein to neurons[53]. The thicker the brain cortex becomes, the further the neuroblasts must migrate[54]. Chemical substances which act on cell migration can prevent formation of the typical layers in different brain areas and cause functional disturbances postnatally. Cyclophosphamide blocked the mitoses of the neuroepithelium in the ventricular layer on the 15th and 16th day p.c. and caused cell degeneration in the mantle layer. Malformations of the brain were visible in the course of the third day after application of cyclophosphamide only. Langman and co-workers[41] examined the proliferative activity in the brain of the mouse on the 15th day p.c. by means of autoradiography. After labelling the cells with [^{3}H]thymidine, they found the greatest proliferative activity in the neuroepithelial layer of the lateral ventricles, in the striatal portion of the brain, and in the posterior parts of the roof of the mesencephalon, near to the cerebellum. After application of 5-azacytidine on the same developmental day, many neuroblasts died in the very locations which were labelled with tritiated thymidine. In later stages of development only (19th day p.c.) definite cell layers, which can be damaged by 5-azacytidine, appeared in the dentate gyrus, the olfactory bulb, and the cerebellum. If these findings are compared with our results, the localization of the cell necroses is very much in agreement, taking into consideration that in early fetogenesis the rat has developed in one day somewhat more than the mouse. The cell necroses induced after application of cyclophosphamide do not coincide with the 'physiological' necroses occurring during normogenesis (for review see refs. 20 and 55).

It is evident that in late pregnancy entodermal derivatives do not respond to cyclophosphamide. Of the ectodermal derivatives, the thoracic spinal cord, all cranial and spinal ganglia, both lobes of the hypophysis, the retina and the lens, the epithelium of the nasal and mouth cavities, the dental lamina, and the labyrinth were always spared. Also the cartilaginous nasal and otic capsule, as described in the literature[56], were free from degeneration.

The anomalous development of the fetus following treatment with cyclophosphamide at the dose level used became evident, macroscopically, in the course of the third day only. The body oedema was more generalized than on the 21st day p.c., by which time large epidermal blebs had developed. Skin blebs, partly filled with blood, were also observed after exposure to induced hypoxia[57]. The skin of the head in the region of the touch hairs responded in a very severe manner. As histological signs, severe degeneration of the mesenchymal sheath of the hair roots was detected, so that the epithelial parts of the hair without mesenchyme traversed the connective tissue cysts. During fetogenesis we have observed that the hair anlages were free of spontaneous degenerations[58]. After degeneration was completed, extensive gelatinous oedemas were present in the eyelids which remained open as a consequence of developmental arrest. Similar observations were made in the mouse after application of sodium arsenate[59] and in the chicken after cyclophosphamide[2]. The structure of the retina was always inconspicuous in our material, though malformations of the eyes induced by cyclophosphamide were frequent in earlier stages[61]; also, postnatal disturbances in the differentiation of the photoreceptors can appear[32].

The necroses induced on the body of the fetuses did not necessarily involve structural developmental defects. The severe cell death in the genital tubercle, without any visible anomalies later on, may serve as an example. On the other hand, degenerative processes were observed in the anlages of the extremities. The latter, above all, occurred in the subcutaneous mesenchyme but not in the condensed cartilaginous and muscular blastemas. In their localization these coincided with 'physiological' necroses appearing at the corresponding developmental stage of the rat (for review see ref. 20). Malformation of the extremities belongs to the embryopathies[16, 20]. The single defect of the footplate, which we observed as oligodactylia after application of cyclophosphamide on the 15th day, belonged to the fetus whose degree of development corresponded to the situation on the 14th day. Abnormal positions of the extremities also appeared without recognizable defects of the cartilaginous and bony skeleton as a result of altered tension between synergists and antagonists of affected muscles. Striking deviations of the skeletal system, mostly retardations, were visible on the 21st day; these appeared as a consequence of the recognizable decrease in weight and reduction in size of 30% of the fetuses on the average, in contrast to the controls. In addition, a decreased deposition of mineral salts would seem to explain the transparency of single bones.

With respect to the affected structure, degenerative and reparative processes did not differ from those which occurred after changes during the embryonic phase of development[17, 20]. Following destruction in the beginning, decomposition and reparation paralleled each other. Finally, only reparation processes were observed.

Besides inhibition of the ventricular mitoses, severe necroses in the mantle layer were observed in the pallium where neuroblasts differentiate to neurons. The disturbances in differentation involved the following patho-histological events: karyopycnosis, karyorhexis, and (seldom) karyolysis. Within the first day after cyclophosphamide damage, numerous basophilic degeneration granules were found scattered in the mantle layer. They consisted of severely condensed nuclear DNA. In the damaged areas, single large cytoplasmic complexes were present which contained the debris of numerous nuclei in an eosinophilic cytoplasmic mass. A vital nucleus cannot be recognized within them with certainty by light microscopic methods. We believe that macrophages are not involved here, but necrotic cells whose cytoplasms have fused because of altered surface tensions. These necrotic conglomerates were PAS-positive but the contents did not contain glycogen.

On the second day after damage, the necrotic regions became more and more infiltrated with oedematous fluid. The blood vessels, already completely filled within the first 24 h, could rupture and form large blood extravasation, especially in the brain and subcutaneously. More and more, macrophages appeared which contained numerous phagocytized nuclear debris besides a normal nucleus. If one did not find a vital nucleus in the histological section, they were similar to the necrotic herds. We doubt whether there is any relationship between the necrotic conglomerates, which appeared shortly after damage, and the cytoplasmatic complexes (fetal macrophages) which appeared later on. The latter are described as 'embryonic waste cells' ('Embryonale Abräumzellen')[26, 35] after induced cell death and is equivalent to the necrotic conglomerates observed by other workers[62-67] in cases of spontaneous cell death during normal development. In previous studies[20], we demonstrated lysosomal enzyme systems (acid phosphatases, esterases) in the macrophages, following treatment of the conceptus during the late embryonic period of development by cyclophosphamide (13th day p.c.; rat). Macrophages were often localized in a small cavity, free from necrotic cell debris which obviously appeared by its activity. At 72 h after application of cyclophosphamide many of the macrophages were less distinct; we believe that they degenerated and were resorbed after having finished their activity. But it is not only macrophages which take part in the removal of necrotic cell debris. The material expelled into the brain ventricles three days after damage is at term no more identifiable: it dissolved within the cerebrospinal fluid. The reparation processes could lead to defective healing or to morphological *restitutio ad integrum*. Whether the latter, especially in the central nervous system, also implies functional integrity, is doubtful. Many noxae induced, during late pregnancy, postnatal behavioural and functional disturbances in the activity of the central nervous system, without any evidence of morphological changes: vitamin A[68, 69], d-amphetamine[70], zinc deficiency[71], and methadone[72]. The human brain also shows increased sensitivity during the last three months of pregnancy and during the postnatal period[73] with respect to later functional performances. Studies with methylazoxymethanol[74], methylmercury[75], and 5-iododeoxyuridine[76] showed that perinatal or postnatal permanent damage of the central nervous system can be induced. Numerous drugs used in the therapy of tumours have toxic, teratogenic, mutagenic, and antimitotic effects in both animal experiments[60, 77-81], and in man[82, 83]. We have shown[84] that following chronic application of cyclophosphamide the teratogenic effect is reduced to a large extent in favour of an embryotoxic one; most likely in this case enzyme induction in the microsomal fraction of the liver is involved. After activation of cyclophosphamide in the liver of the mother[85], the drug rapidly passes across the placenta to the fetus and produces multilayered effects in the cycle of embryonic cells (for review see refs. 86–89). Individual authors[55, 90] considered the influence of cyclophosphamide on spontaneous necroses during normogenesis, the substance intensifying the normal areas of degeneration from which irreversible damage results. We did not observe such effects during fetal development.

ACKNOWLEDGEMENTS

I wish to thank especially Prof. Dr. sc. med. R. Bertolini, Head of the Department of Anatomy of the Karl-Marx-University of Leipzig, for his valuable assistance, and Mrs P. Wendler, medical assistant, for preparing the histological material and the photographs.

References

1. Gerlinger, P. (1966). Contribution à l'étude de l'action tératogène du cyclophosphamide chez l'embryon de poulet et de lapin. *Arch. Anat. Histol. Embryol.*, 49, 249
2. Gerlinger, P., Ruch, J. V. and Clavert, J. (1962). Note préliminaire sur l'action du cyclophosphamide (endoxan) sur le développement de l'embryon. *C. R. Acad. Sci.*, 255, 3229
3. Gerlinger, P., Ruch, J. V. and Clavert, J. (1963). Action du cyclophosphamide sur la formation de l'embryon. *C. R. Séan. Soc. Biol.*, 157, 173
4. Miclea, C. and Arcan, A. (1967). Effects of clafen (cyclophosphamide) injected into the ependymal canal of the chick embryo. *Rev. Roum. Embryol. Cytol. Sér. Embryol.*, 4, 163
5. Gebhardt, D. O. E. (1970). The embryolethal and teratogenic effects of cyclophosphamide on mouse embryos. *Teratology*, 3, 273
6. Gibson, J. E. and Becker, B. A. (1967). Teratogenicity of cyclophosphamide in the mouse. *Toxicol. Appl. Pharmacol.*, 10, 380
7. Gibson, J. E. and Becker, B. A. (1968). The teratogenicity of cyclophosphamide in mice. *Cancer Res.*, 28, 475
8. Hackenberger, I. and Kreybig, Th.v. (1965). Vergleichende teratologische Untersuchungen bei der Maus und bei der Ratte. *Arzneimittelforsch.*, 15, 1456
9. Schmidt, W., Kreutz, R., Wendler, D. and Gabler, W. (1977). Entstehung von Skelettschäden nach Verabreichung von Aminoacetonitril und Cyclophosphamid während der Fetogenese der Ratte. *Anat. Anz.*, 142, 635
10. Ashby, R., Davis, L., Dewhurst, B. B., Espinal, R., Penn, R. N. and Upshall, D. G. (1976). Aspects of the teratology of cyclophosphamide (NSC-26271). *Cancer Treat. Rep.*, 60, 477
11. Botta, J. A., Hawkins, H. C. and Wikel, J. H. (1974). Effects of cyclophosphamide on fertility and general reproductive performance of rats. *Toxicol. Appl. Pharmacol.*, 27, 602
12. Brock, N. and Kreybig, Th.v. (1964). Experimenteller Beitrag zur Prüfung teratogener Wirkungen von Arzneimitteln an der Laboratoriumsratte. *Naunyn-Schmiedebergs Arch. Exp. Pathol. Pharmakol.*, 249, 117
13. Chaube, S., Kury, G. and Murphy, M. L. (1967). Teratogenic effects of cyclophosphamide (NSC-26271) in the rat. *Cancer Chemother. Rep.*, 51, 363
14. Kreybig, Th.v. (1965). Zur Wirkung von Teratogenen auf frühe Stadien der vorgeburtlichen Entwicklung der Ratte. *Naunyn-Schmiedeberg's Arch. Exp. Pathol. Pharmakol.*, 252, 196
15. Kreybig, Th.v. (1965). Die teratogene Wirkung von Cyclophosphamid während der embryonalen Entwicklungsphase bei der Ratte. *Naunyn-Schmiedeberg's Arch. Exp. Pathol. Pharmakol.*, 252, 173
16. Kreybig, Th.v. (1968). *Experimentelle Praenatal – Toxikologie*. (Aulendorf i. Württ.: Editio Cantor KG)
17. Kreybig, Th.v. and Schmidt, W. (1966). Zur chemischen Teratogenese bei der Ratte. *Arzneimittelforsch.*, 16, 989
18. Singh, S. and Sanyal, A. K. (1974). Skeletal malformations of forelimbs of rat fetuses caused by maternal administration of cyclophosphamide during pregnancy. *J. Anat.*, 117, 179
19. Tuchmann-Duplessis, H. (1969). The action of anti-tumour drugs on gestation and on

embryogenesis. In: A. Bertelli and L. Donati (eds.). *Teratology.* Excerpta Medica Int. Congr. Ser. 173, pp. 75–86. (Amsterdam: Excerpta Med. Found.)
20. Wendler, D. (1972). Der embryofetale Zelltod während der Normogenese und im Experiment. Acta histor. Leopoldina, NF 8, pp. 1–295. (Leipzig: Barth)
21. Wilson, J. G. (1964). Teratogenic interaction of chemical agents in the rat. *J. Pharmacol. Exp. Ther.*, **144**, 429
22. Fritz, H. and Hess, R. (1971). Effects of cyclophosphamide on embryonic development in the rabbit. *Agents Actions*, **2**, 83
23. Gerlinger, P. and Clavert, J. (1965). Action du cyclophosphamide injecté à différentes périodes de la gestation sur les cellules sexuelles embryonnaires de lapin. *C. R. Séan. Soc. Biol.*, **158**, 2464
24. Gottschewski, G. H. M. and Zimmermann, W. (1963). Auslösung von Blastopathien beim Säugetier durch Cyclophosphamid und Thalidomid. *Naturwissenschaften*, **50**, 525
25. Gottschewski, G. H. M. and Zimmermann, W. (1964). Nachweis von phänokritischen Phasen verschiedener Organanlagen beim Hauskaninchen (Oryctolagus cuniculus). *Zool. Anz.*, **27** (Suppl.), 144
26. Kreybig, Th.v. and Schmidt, W. (1967). Chemisch induzierte Fetopathien bei der Ratte. Experimentelle Untersuchungen über die Wirkung von Cyclophosphamid und N-Methyl-N-nitroso-Harnstoff nach der Gabe am 15. oder 16. Tag der Gestation. *Arzneimittelforsch.*, **17**, 1093
27. Nishimura, H. and Tanimura, T. (1976). *Clinical Aspects of the Teratogenicity of Drugs.* (Amsterdam–Oxford–New York: Excerpta Medica, American Elsevier Publ. Comp. Inc.)
28. Sieber, S. M. and Adamson, R. H. (1975). Toxicity of antineoplastic agents in man: chromosomal aberrations, antifertility effects, congenital malformations, and carcinogenic potential. *Adv. Cancer Res.*, **22**, 57
29. Nordlinder, H. (1971). Malformations in newborn rats treated with a single dose of cyclophosphamide. *Acta Soc. Med. upsaliensis*, **76**, 87
30. Vahlsing, H. L., Feringa, E. R., Britten, A. G. and Kinning, W. K. (1975). Dental abnormalities in rats after a single large dose of cyclophosphamide. *Cancer Res.*, **35**, 2199
31. Stekar, J. (1973). Teratogenicity of cyclophosphamides in newborn rats. *Arzneimittelforsch.*, **23**, 922
32. Foerster, H. and Lierse, W. (1975). Vulnerabilität früher postnataler Differenzierungsvorgänge der Retina und teratogener Effekt des Zyklophosphamids (Endoxan). *Acta Anat.*, **93**, 161
33. Nordlinder, H. (1969). Malformations after treatment of new-born mice with a single dose of cyclophosphamide. *Experientia*, **25**, 1296
34. Bus, J. S. and Gibson, J. E. (1973). Teratogenicity and neonatal toxicity of isofamide in mice. *Proc. Soc. Exp. Biol. Med.*, **143**, 965
35. Kreybig, Th.v. (1969). The critical sensitivity of the developmental phase and the organotropic action of different teratogenic agents; receptors of morphogenesis in the mammalian embryo. In: A. Bertelli and L. Donati (eds.). *Teratology.* Int. Congr. Ser. No. 173, pp. 152–159. (Amsterdam: Excerpta Medica Foundation)
36. Koppang, H. S. (1973). Histomorphologic investigations on the effect of cyclophosphamide on dentinogenesis of the rat incisor. *Scand. J. Dent. Res.*, **81**, 383
37. Singh, S. and Singh, M. (1975). Histochemical changes in the malformed forelimbs of rat fetuses induced by cyclophosphamide. *J. Anat. Soc. India*, **24**, 53
38. Salzgeber, B. (1966). Production élective de la phocomélie sous l'influence d'ypérite azotée, chez l'embryon de poulet. II. Etude histologique des bourgeons de membres au cours du développement. *J. Embryol. Exp. Morphol.*, **16**, 339
39. Kreybig, Th.v., Preussmann, W. and Schmidt, W. (1968). Chemische Konstitution und teratogene Wirkung bei der Ratte. I. Carbonsäureamide, Carbonsäurehydrazide und Hydroxamsäuren. *Arzneimittelforsch.*, **18**, 645
40. Desesso, J. M. (1975). A comparative light microscopic study of the teratogenic effects of three drugs on rabbit limb development. *Anat. Rec.*, **181**, 345
41. Langman, J., Webster, W. and Rodier, P. (1975). Morphological and behavioural abnormalities caused by insults to the CNS in the perinatal period. In: C. L. Berry and D. E. Poswillo (eds.). *Teratology. Trends and Applications*, pp. 182–200. (Berlin–Heidelberg–New York: Springer)

42. Barlow, S. M. and Sullivan, F. M. (1975). Behavioural teratology. In: C. L. Berry and D. E. Poswillo (eds.). *Teratology. Trends and Applications*, pp. 103–120. (Berlin–Heidelberg–New York: Springer)
43. Coyle, I., Wayner, M. J. and Singer, G. (1976). Behavioral teratogenesis: A critical evaluation. *Pharmacol. Biochem. Behav.*, **4**, 191
44. Wendler, D., Gabler, W., Schmidt, W. and Pabst, R. (1976). Wirkungsspezifität chemischer Teratogene nach Abschluß der Organogenese. *Anat. Anz.*, **140**, 405
45. Chaube, S. and Murphy, M. L. (1969). Fetal malformations produced in rats by N-isopropyl-α-(2-methylhydrazino)-p-toluamide hydrochloride (procarbazine). *Teratology*, **2**, 23
46. Percy, D. H. (1975). Teratogenic effects of the pyrimidine analogs 5-iododeoxyuridine and cytosine arabinoside in late fetal mice and rats. *Teratology*, **11**, 103
47. Barr, M. and Brent, R. L. (1970). The relation of the uterine vasculature to fetal growth and the intrauterine position effect in rats. *Teratology*, **3**, 251
48. DeMyer, W. and Baird, I. (1969). Mortality and skeletal malformations from amniocentesis and oligohydramnios in rats: Cleft palate, clubfoot, microstomia, and adactyly. *Teratology*, **2**, 33
49. Brent, R. L. (1969). The direct and indirect effects of irradiation upon the mammalian zygote, embryo, and fetus. In: H. Nishimura and J. R. Miller (eds.). *Methods for Teratological Studies in Experimental Animals and Man*, pp. 63–75. (Tokyo: Igaku Shoin Ltd.)
50. Halfpap, E. (1962). Embryo- und Fetopathien, unter besonderer Berücksichtigung parasitärer Infektionen. *Zbl. Gynäkol.*, **37**, 1437
51. Wendler, D. (1977). Die experimentelle Teratologie im Dienste der teratologischen Routineprüfung. *Anat. Anz.*, **142**, 597
52. Chatel, M. (1976). Développement de l'isocortex du cerveau humain pendant les périodes embryonnaires et foetales jusqu'à la 24 ème semaine de gestation. *J. Hirnforsch.*, **17**, 189
53. Langman, J. and Welsh, G. W. (1967). Excess vitamin A and development of the cerebral cortex. *J. Comp. Neurol.*, **131**, 15
54. Berry, M. and Rogers, A. W. (1965). The migration of neuroblasts in the developing cerebral cortex. *J. Anat.*, **99**, 691
55. Menkes, B., Sandor, S. and Ilies, A. (1970). Cell death in teratogenesis. In: D. H. M. Woollam (ed.). *Advances in Teratology*, 4, pp. 169–215. (London: Logos Press)
56. Schreiner, L. and Kreybig, Th.v. (1968). Chemisch erzeugte Mißbildungen im Tierexperiment aus klinischer Sicht. *Arch. Klin. Exp. Ohren-, Nasen- Kehlkopfheilkd.*, **190**, 330
57. Grabowski, C. T. (1970). Embryonic oxygen deficiency – a physiological approach to analysis of teratological mechanisms. In: D. H. M. Woollam (ed.). *Advances in Teratology*, **4**, pp. 207–234. (New York–London: Academic Press)
58. Noack, W. and Schweichel, J.-U. (1971). Elektronenmikroskopische Untersuchungen des mesenchymwärts gerichteten Epithelwachstums während der frühen Entwicklung von Vibrissae bei Rattenembryonen. *Anat. Anz.*, **130**, 297
59. Matsumoto, N., Okino, T., Katsunuma, H. and Iijima, S. (1973). Effects of sodium arsenate on the growth and development of the fetal mice. *Teratology*, **8**, 98
60. Hemsworth, B. N. and Jackson, H. (1965). Embryopathies induced by cytotoxic substances. In: J. M. Robson, F. M. Sullivan and R. L. Smith (eds.). *Symposium on Embryopathic Activity of Drugs*, pp. 116–137. (London: Churchill Livingstone)
61. Singh, S. and Sanyal A. K. (1976). Eye anomalies induced by cyclophosphamide in rat fetuses. *Acta Anat.*, **94**, 490
62. Ernst, M. (1926). Über Untergang von Zellen während der normalen Entwicklung bei Wirbeltieren. *Z. Anat. Entw.-Gesch.*, **79**, 228
63. Glücksmann, A. (1930). Über die Bedeutung von Zellvorgängen für die Formbildung epithelialer Organe (Linse, Augenblase, Neuralrohr usw.). *Z. Anat. Entw.-Gesch.*, **93**, 35
64. Graumann, W. (1950). Zelldegeneration im Telencephalon medium und Paraphysenentwicklung bei der weißen Maus. *Z. Anat. Entw.-Gesch.*, **115**, 19
65. Hinrichsen, K. (1959). Über das Vorkommen von Desoxyribonucleinsäure im Cytoplasma embryonaler Epithelzellen. *Anat. Anz.*, **106/107 (Suppl.)**, 112
66. Peter, K. (1936). Untersuchungen über Zelluntergang in der Embryogenese. *Z. Anat. Entw.-Gesch.*, **105**, 409

67. Theiler, K. (1958). Zelluntergang in den hintersten Rumpfsomiten bei der Maus. *Z. Anat. Entw.-Gesch.*, **120**, 274
68. Hutchings, D. E. and Gaston, J. (1974). Effects of vitamin A excess administered during the mid-fetal period on learning and development in rat offspring. *Dev. Psychobiol.*, **7**, 225
69. Hutchings, D. E., Gibbon, J. and Kaufman, M. A. (1973). Maternal vitamin A excess during the early fetal period effects on learning and development in the offspring. *Dev. Psychobiol.*, **6**, 445
70. Seliger, D. L. (1975). Prenatal maternal D-amphetamine effects on emotionally and audiogenic seizure susceptibility of rat offspring. *Dev. Psychobiol.*, **8**, 261
71. McKenzie, J. M., Fosmire, G. J. and Sandstead, H. H. (1975). Zinc deficiency during the latter third of pregnancy effects on fetal rat brain, liver, and placenta. *J. Nutr.*, **105**, 1466
72. Hutchings, D. E., Hunt, H. F., Towey, J. P., Rosen, T. S. and Gorinson, H. S. (1976). Methadone during pregnancy in the rat: dose level effects on maternal and perinatal mortality and growth in the offspring. *J. Pharmacol. Exp. Ther.*, **197**, 171
73. Wiglesworth, J. S. (1968). Disorders of fetal growth. *J. Obstet. Gynaecol. Br. Commonw.*, **75**, 1234
74. Shimada, M. and Langman, J. (1970). Repair of the external granular layer of the hamster cerebellum after prenatal and postnatal administration of methylazoxymethanol. *Teratology*, **3**, 119
75. Sobotka, T. J., Cook, M. P. and Brodie, R. E. (1974). Effects of perinatal exposure to methylmercury on functional brain development and neurochemistry. *Biol. Psychiatry*, **8**, 307
76. Percy, D. H. and Albert, D. M. (1974). Developmental defects in rats treated postnatally with 5-iododeoxyuridine (IUDR). *Teratology*, **9**, 275
77. Adams, C. E., Hay, M. F. and Lutwak-Mann, C. (1961). The action of various agents upon the rabbit embryo. *J. Embryol. Exp. Morphol.*, **9**, 468
78. Chaube, S. and Murphy, M. L. (1968). The teratogenic effects of the recent drugs active in cancer chemotherapy. In: D. H. M. Woollam (ed.). *Advances in Teratology*, **3**, pp. 181–237. (London: Logos Press)
79. Connors, T. A. (1975). Cytotoxic agents in teratogenic research. In: C. L. Berry and D. E. Poswillo (eds.). *Teratology. Trends and Applications*, pp. 49–79. (Berlin–Heidelberg–New York: Springer)
80. Degenhardt, K. H. and Kleinebrecht, J. (1971). Teratogenic effects of antitumor drugs. In: H. Tuchmann-Duplessis (ed.). *Malformations Congénitales des Mammifères*, pp. 79–94. (Paris: Masson)
81. Murphy, M. L. (1960). Teratogenic effects of tumour-inhibiting chemicals in the foetal rat. In: G. E. W. Wolstenholme and C. M. O'Connor (eds.). *Ciba Foundation Symposium on Congenital Malformations*, pp. 78–107. (London: Churchill Livingstone)
82. Greenberg, L. H. and Tanaka, K. R. (1964). Congenital anomalies probably induced by cyclophosphamide. *J. Am. Med. Assoc.*, **188**, 423
83. Toledo, T. M., Harper, R. C. and Moser, R. H. (1971). Fetal effects during cyclophosphamide and irradiation therapy. *Ann. Intern. Med.*, **74**, 87
84. Pabst, R. and Wendler, D. (1976). Zur Bewertung der chronischen Arzneimittelapplikation in der Teratologie. *Anat. Anz.*, **140**, 413
85. Brock, N. and Hohorst, H.-J. (1962). Über die Aktivierung von Cyclophosphamid im Warmblütlerorganismus. *Naturwissenschaften*, **49**, 610
86. Druckrey, H., Steinhoff, D., Nakayama, M., Preussmann, R. and Anger, K. (1964). Beiträge zum Dosisproblem in der Krebs-Chemotherapie und zur Wirkungsweise des Endoxans. In: H. Wilmanns (ed.). *Chemotherapie Maligner Tumoren*, pp. 11–36. (Stuttgart: Schattauer)
87. Lawley, P. D. and Brookes, P. (1963). The action of alkylating agents on deoxiribonucleic acid in relation to biological effects of the alkylating agents. *Exp. Cell Res.*, **9**, 512
88. Loveless, A. (1966). *Genetic and Allied Effects of Alkylating Agents*. (London: Butterworths)
89. Ross, W. C. J. (1962). *Biological Alkylating Agents*. (London: Butterworths)
90. Pexieder, T. (1974). Der Einfluß von Cyclophosphamid auf die physiologischen Zelltodzonen im Herz des Hühnchens. *Anat. Anz.*, **136**, 841

7
The effect of retinoic acid on the developing hamster heart — an ultrastructural and morphological study

I. M. TAYLOR

INTRODUCTION

In 1956, Fox and Goss[1] suggested that the administration of teratogens such as trypan blue to developing mammalian embryos might afford a new method of experimental attack on the problems of normal morphogenesis of the heart. Sadly, however, Van Mierop and Gessner[2] 16 years later concluded that it had not so far been possible either by drug or mechanical interference to induce specific anomalies reproducibly in a high enough percentage of offspring to render such studies either practical or reliable. On a more positive note, they did feel that studies such as those of Patterson[3] and Siller[4] who have examined dogs and chickens with genetic disposition to certain cardiac defects had gone some way towards the goal of understanding the various aetiologies and mechanisms underlying these congenital abnormalities.

As indicated, many methods have been used to try and produce heart defects with varying degrees of success (for review see Pexieder[5]). For example, several have attempted to produce bulboventricular defects by the use of cyclophosphamide, dexamethasone and thalidomide[5-7]. Their results, and those of other workers, generally indicate that such cardiac defects may be found in as many as 90% of survivors, although there is a very high fetal mortality. Despite the fact that many fetuses have the same defect such figures are not ideal. The mortality rates imply that study of abnormal cardiogenesis using fetuses obtained by these methods is going to be imprecise at best, not least because it is so difficult to predict at any point during development if any one fetus would have a specific cardiac anomaly if it were to be examined at term should it survive that long.

A better cardiac teratogen would, therefore, result in a much lower fetal mortality combined with a similarly high frequency of cardiac defects in treated offspring.

Vitamin A and related compounds have occasionally been observed to have a teratogenic effect on the heart and great vessels of several mammalian species[8-12]. Shenefelt[12] in particular has reported that 12% of the offspring of hamsters treated with retinoic acid relatively early in organogenesis had transposition of the great vessels and many others had ventricular septal defects. The malformations were discovered by means of razor blade sectioning of near-term fetuses. However, recent experience suggests that this method is less than ideal for detecting cardiac defects. It was, therefore, decided to re-examine this finding using microdissection in the hope that a much higher percentage of abnormal hearts could be detected at term[13].

In addition, the gross and ultrastructural appearances of the fetal hamster heart have been studied at various times after maternal administration of retinoic acid.

MATERIALS AND METHODS

Golden Syrian hamsters of approximately 110 g, obtained from Trenton Experimental Laboratory, were mated from 7 to 9 p.m. and the following day was designated as day 0 of gestation. Mated animals were kept singly in wire cages at ambient temperature in a room with controlled light and dark periods of 12 h each. Purina Lab Chow and tap water were provided *ad libitum*.

Mated animals were divided into three groups: untreated controls, corn oil controls which received a single dose of a corn oil vehicle, and retinoic acid treated animals which received a single dose of retinoic acid suspended in corn oil. All injections were by gavage and each animal received only one injection. The effect of treatment on each of days 7 to 10 was tested. A dose of 80 mg/kg of retinoic acid maternal body weight was given; however, because a high rate of fetal mortality resulted from this dose level on day 7 of gestation, the effects of 40 and 60 mg/kg on this day were also tested. The females were killed by prolonged chloroform anaesthesia on the evening of day 14 and the uterine contents examined. The numbers of dead and resorbed fetuses were noted and the survivors fixed in formol saline. The chests were opened and the heart, great vessels and lungs removed *en bloc*. Following examination of the surface features of the hearts they were then carefully dissected in order to examine their internal structure.

Three further groups of animals which were treated on day 8 of gestation only were killed on days 9, 10, 11, 12 or 13. Those fetuses which were living were prepared as before and examined in identical fashion for cardiac defects.

The hearts from offspring of similar groups of animals were also examined by electron microscopy. They were obtained 2, 4, 6, 8, 12, 16, 18 or 24 h after maternal treatment and only those hearts which were beating were processed. All specimens were fixed for 1 h in cold 2% paraformaldehyde and 2% glutaraldehyde in 0.1 M phosphate buffer at pH 7.4. After secondary fixation in 1% osmium tetroxide, they were dehydrated in alcohol and embedded in Epon 812. Thin sections were stained with lead citrate and uranyl acetate and examined in a Philips 300 electron microscope.

RESULTS

The detailed results from the first part of this study will be published elsewhere, and only those referring to the bulboventricular malformations will be mentioned here.

Clearly retinoic acid had a profound effect on the development of the heart and great vessels of hamster fetuses exposed to this teratogen early in organogenesis (Table 7.1).

Table 7.1 The effects of retinoic acid on the hearts and great vessels of golden Syrian hamster fetuses from litters exposed to single doses by gavage during gestation and examined on day 14

Day of treatment	Dose (mg/kg)	No. females	No. with complete litter loss	Implants*	Mortality† %	No. examined/ live	Abnormalities‡ %
Untreated		7	0	82	2.4	65/80	0.0
7	Control	5	0	57	1.8	48/56	0.0
	40	12	1	119	28.6	78/85	33.3
	60	16	7	119	36.1	69/76	39.1
	80	13	9	46	67.4	13/15	69.1
8	Control	3	0	30	10.0	27/27	3.7
	80	18	6	135	33.3	82/90	74.4
9	Control	7	0	80	10.0	70/72	0.0
	80	12	2	100	7.0	83/93	30.1
10	Control	8	0	94	5.3	83/89	1.1
	80	14	3	114	21.1	86/90	12.8

* The figures represent the number of implantation sites in females with at least one surviving fetus at term
† The values are expressed as per cent of implants
‡ The values are expressed as per cent of the hearts examined

The highest rates of heart malformation resulted from maternal treatment on days 7, 8 and 9 of pregnancy. Exposure on day 8 resulted in 74% of surviving fetuses with detectable cardiovascular deformities while administration of the teratogen on day 9 produced cardiac defects in only 30.1%.

The most frequently seen abnormalities were isolated ventricular septal defects (VSD), double outlet right ventricle (DORV), transposition of the great vessels and an overriding aorta complex. Isolated defects of both the muscular and the membranous parts of the interventricular septum occurred in consistently moderate frequencies on all days tested, however, the VSDs observed following treatment on gestation day 8 most often accompanied transposition or DORV. Maternal treatment on day 8 of gestation resulted in 11% of the surviving offspring having complete transposition of the great vessels. In these fetuses the aorta was in continuity with the morphological right ventricle and at least half of the pulmonary trunk arose from a morphological left ventricle. The most commonly seen abnormality was double outlet right ventricle, which occurred in 46% of fetuses surviving exposure on day 8 of gestation (Table 7.2). The hearts of these fetuses had at least $1\frac{1}{2}$ great vessels arising from the morphological right ventricle in association with discontinuity between the mitral and semilunar valves. The relationship of the

Table 7.2 Frequencies* of specific abnormalities in the hearts and great vessels of hamster fetuses from females treated during gestation with single doses of retinoic acid and killed on day 14

	Dose (mg/kg)										
	Day 7				Day 8		Day 9		Day 10	
	Control	40	60	80	Control	80	Control	80	Control	80
Ventricular septal defect	0.0	17.9	26.1	15.4	3.7	22.0	0.0	16.9	1.2	12.8
Transposition	0.0	1.3	5.8	7.7	0.0	11.0	0.0	2.4	0.0	2.3
Double outlet right ventricle	0.0	2.6	1.5	7.7	0.0	46.3	0.0	3.6	0.0	0.0
Overriding aorta in association with VSD and pulmonary stenosis	0.0	1.3	1.5	7.7	0.0	4.9	0.0	9.6	0.0	3.5

* The frequencies are expressed as percent of hearts examined as given in Table 7.1

great vessels to each other, however, was variable. Most frequently, the aorta was situated to the right and anterior to the pulmonary trunk, although the normal relationship as well as intermediate stages between these two extremes were observed. Hearts with overriding aorta and mitral–aortic continuity in association with a ventricular septal defect and pulmonary stenosis could be found after treatment on any day although the highest frequency (9.6) followed maternal exposure on day 9 of pregnancy.

The results which follow are preliminary and necessarily incomplete. However, gross study of intact hearts reveals that retinoic acid administered to the mother affects the morphogenesis of the fetal heart very rapidly. For example, after treatment on day 7, some treated hearts are clearly different from control specimens within 2–6 h. The heart tube is distorted and the bulbus and ventricle appear shorter than those of normal fetuses. Furthermore, this part of the heart has an abnormal spatial relationship to the atrioventricular canal and atrium.

In older specimens there is obvious general retardation of growth and gross structural abnormalities are identifiable. On day 12, for example, hearts exhibiting the three major bulboventricular malformations mentioned earlier can be detected without difficulty (Figures 7.1 and 7.2).

There are marked changes in the ultrastructural appearance of cells in both the ventricular and bulbar myocardium as well as in both the bulbar cushions. However, only minor effects such as cell budding have been observed in cells of the endocardium.

Two hours after administration of drug to the mother there are already abnormal findings in both the myocardial and mesenchymal cells of fetuses no matter on which day the treatment was given. Swelling of the plasma membrane associated with budding of cells, dilatation of the rough endoplasmic reticulum, and accumulation of lipid droplets are the most obvious early signs during the first six hours (Figure 7.3). Areas where cytoplasmic organization is lost appear together with both intracellular and extracellular vesicles containing flocculent material (Figure 7.4). There is also increasing separation of cells and an apparent enlargement of the extracellular space. These findings are most noticeable on days 7 and 8, to a lesser extent on day 9 and least on day 10.

After 12–24 h, dying and dead cells are a prominent feature of the bulboventricular region. Although such cells are found normally, it should be emphasized that retinoic acid treated hearts exhibited far more death than control specimens in any part of the heart which has been examined so far. Myocardial cells lose their polyribosomes, develop nuclear membrane irregularities and fragment (Figure 7.5). Their nuclei become abnormal, most appear very pale while others become pyknotic and are then extruded (Figure 7.6). Such dead and dying cells may be found in groups throughout the bulboventricular myocardium (Figure 7.7) but they are especially found in or near the bases of the proximal bulbar cushions. Similarly, some mesenchymal cells lose their ground cytoplasm, develop swollen or condensed mitochondria and die.

In specimens 12–24 h after maternal retinoic acid administration it can be seen that portions of degenerating cells have been phagocytosed by healthy

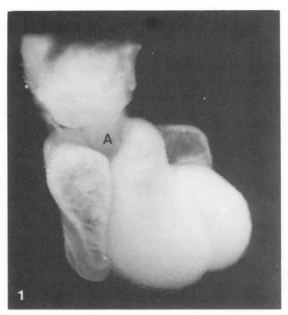

Figure 7.1 A normal day 12 fetal heart viewed from the right ventrolateral aspect. Note that the pulmonary trunk arises ventrally and to the right of the aorta A

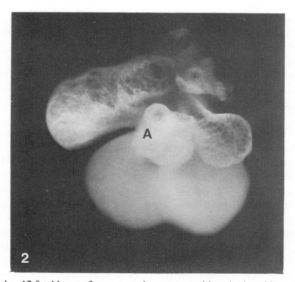

Figure 7.2 A day 12 fetal heart after maternal treatment with retinoic acid on day 8. A double outlet right ventricle is clearly seen, the aorta A in this specimen arises lateral to the pulmonary trunk

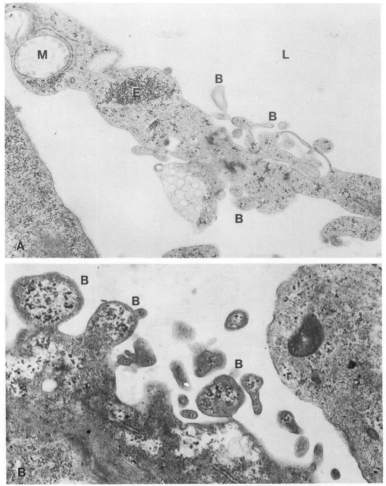

Figure 7.3 A, Budding B and swelling of mitochondria M are seen in the endothelial cell E. Budding is found on both the luminal L and myocardial aspects of the cell 24 h after treatment with Retinoic acid on day 8. × 11 172. B, Profuse budding B is observed on the luminal aspect of a developing myocyte 4 h after treatment with retinoic acid on day 8. × 16 188

neighbouring myocardial and mesenchymal cells (Figure 7.8). By 24 h many of the dead cells have been removed and no traces remain in some areas except for the phagocytotic vacuoles.

The retinoic acid treated hearts seem to fall about 12 h behind control hearts in terms of both their general appearance and their myofibrillar development. However, 24 h after treatment they regain such features as profuse glycogen and abundant rough endoplasmic reticulum which are typically found in the fetal heart.

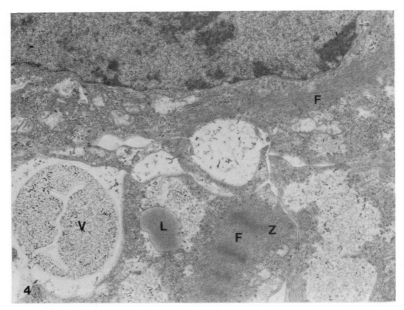

Figure 7.4 Much of the glycogen normally found in the developing myocyte has disappeared although a lipid droplet L, a few organized filaments F and some associated Z band material can be seen. A vesicle V containing flocculent material is found in the extracellular space. Day 9, 2 h after treatment with retinoic acid. × 11 172

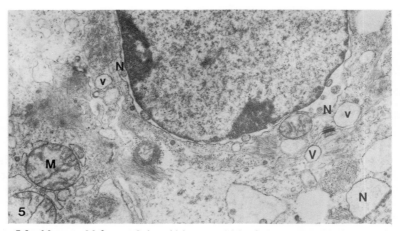

Figure 7.5 Myocyte M from a 9-day-old hamster 16 h after treatment. Nuclear membrane irregularities (N) are observed together with swollen mitochondria M and numerous vacuoles V. A few myofilaments F are also seen. × 11 172

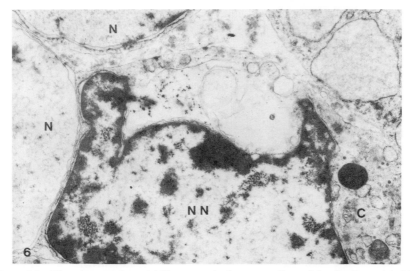

Figure 7.6 Three large, pale nuclei N are seen in the myocardium of the distal ventricle 24 h after treatment on day 10. One of them NN is about to be lost from its cell C. × 11 172

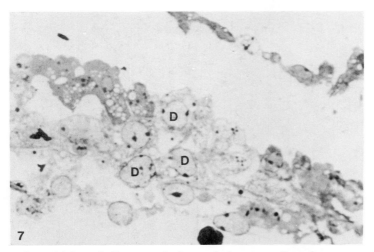

Figure 7.7 Several dying cells D are seen in the ventricular myocardium after treatment 24 h earlier on day 8

DISCUSSION

The results clearly show that retinoic acid is a potent cardiac teratogen in hamsters. Although only 74.4% of the total live fetuses which were treated on day 8 can be shown on day 14 to have an abnormal heart, this is higher than the frequencies found following the use of many other teratogens. At the same time the mortality rates on days 8, 9 and 10 are much lower than those found after using the teratogens identified in the introduction. Moreover, no less

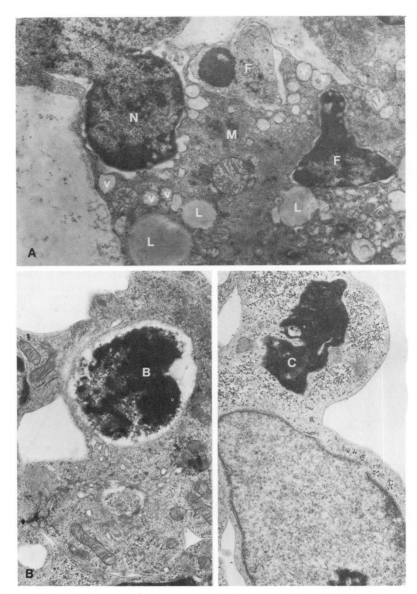

Figure 7.8 A, Twelve hours after treatment on day 9, a myocardial cell M is found in process of phagocytosing a nucleus N. Two other fragment profiles F are observed in the same cell. Note also the lipid droplets L and the many small vacuoles V which seem to be shrunken mitochondria. × 11 172. B and C show later stages in phagocytotic vacuoles found in a myocardial cell B and in a mesenchymal cell C: B, 16 h after treatment on day 9. C, 24 h after treatment on day 8. Both × 11 172

than 87% of this group of live animals, in which an abnormality can be demonstrated, possess a bulboventricular malformation.

These malformations include double-outlet right ventricle and transposition which are found in 62% and 14% respectively, the remaining 11% consisting of the overriding aorta complex. It may, therefore, be said that retinoic acid does produce a majority of cardiac malformations that are reproducible in type. Although Van Mierop and Gessner[2] reported in 1972 that it had not so far been possible to induce any specific anomaly reproducibly, it appears that their conclusion may no longer be true. Analysis of the day 8 treated litters shows that 40% of litters had fetuses with only the double-outlet right ventricle, 50% had both double-outlet right ventricle and transposition, and 10% had double-outlet right ventricle, transposition and the overriding aorta complex. Thus their forecast that embryological studies of developing abnormal hearts would be impractical and unreliable may no longer be valid.

In earlier work, vitamin A and several related compounds have been administered to monkeys, mice, rats, hamsters and guinea pigs with varied results.

In 1977, Fantel et al.[11] found transposition of the great vessels in one of 11 pigtail monkeys given a daily oral dose of 10 mg/kg retinoic acid on days 20 through 44 of gestation.

Kalter and Warkany[8] examined the effect on inbred mice of a single oral dose of 10000 I.U. of vitamin A palmitate on day $8\frac{1}{3}$. They could not demonstrate any difference in response between A/J, DBA/1J and C3H/J mice so far as external appearances were concerned. They found cardiac abnormalities which included transposition, double-outlet right ventricle, overriding aorta and ventricular septal defects. These malformations were revealed by serial 20 μm sections of ten embryos which were severely affected members of those 99 young which were alive at $17\frac{1}{3}$ days after conception. However, no details of the frequency of the defects were published.

Nolan[9] observed the occurrence of ventricular septal defects and ventricular hyperplasia in three strains of rats following oral treatment with 75000 I.U. of vitamin A palmitate on each of days 6 to 15 of pregnancy. Septal defects were found only in Sprague-Dawley rats (4.5%) and ventricular hyperplasia was found solely in Charles River rats (2.8%) while Wistar rats showed neither abnormality. The types and frequencies of these defects were assessed by the Wilson razor blade technique.

Robens[10] investigated the effect of giving either 200 or 400000 I.U. of vitamin A palmitate per kg body weight to hamsters orally on either one or all of days 6, 7, 8, 9 or 10 of gestation. Hypoplastic left heart and persistent truncus malformations were observed together with other unspecified cardiac defects. She also mentions that a guinea pig developed a heart malformation after a single dose of vitamin A palmitate on the 17th day of pregnancy.

Shenefelt[12] treated hamsters with a single oral dose of sodium retinoate (116 mg/kg body weight) on one of days 6 to 9. Following razor blade sectioning, he found several cardiac malformations such as transposition, aortic or pulmonary hypoplasia and ventricular septal defect. The peak frequencies at which these defects were detected were 12%, 28%, 12% and unspecified respectively.

Thus, whereas obvious external features of vitamin A treated fetuses have been described and enumerated in detail, careful study of the cardiac defects has been the exception rather than the rule. Some authors describe individual malformations without mentioning frequency, others mention only a few defects which would seem from the other literature to be unrepresentative of the range of defects which vitamin A is known to produce and some defects are referred to in only the vaguest style. Generally, the frequency of reported defects is rather low and it is, therefore, suggested that the methods by which they have been detected are suspect.

While paraffin sectioning is of proven value in the assessment of cardiac malformations, the use of razor blade sectioning is questionable because many of the defects in the already small fetal hearts are even smaller and thus easily missed because of the plane or thickness of section. However, this is not to deny the worth of this technique in other areas. Although it is not claimed that every cardiac malformation will be found or accurately described following microdissection, the frequency of defects is much higher when this method is used rather than the razor blade technique. A high frequency of abnormalities is essential to the success of cardiac teratology and any method which may unnecessarily reduce the detection rate is thus a major drawback. While it is possible that the differences in both frequency and type of malformations seen in the present study when compared with earlier work may be due in part to the use of retinoic acid itself rather than its sodium salt or vitamin A palmitate, it is felt that it also reflects the advantages of careful inspection and microdissection.

Preliminary study of embryos removed at short intervals after administration of retinoic acid has shown that it is quite easy to pick out the abnormal hearts very early on in their development. This is true even of very young fetuses, as presumptive normal hearts from control animals may be distinguished from abnormal ones within 6 h of treatment. The selective culling of embryos at these and later stages will make it possible to obtain a series of hearts which would almost certainly end up with major bulboventricular malformations if left undisturbed until day 14.

This is opportune, for up to the present there is a lack of agreement concerning the pathogenesis of bulboventricular malformations although they are no longer considered to represent simple arrests of development at different stages. Rather, they are believed by many to be due to maldevelopment of the conus reflecting different combinations of excessive, normal or reduced conal rotation and absorption and of septal migration[14-16]. Others, such as Van Mierop and Wiglesworth[17], hold that abnormal fusion of malformed conotruncal ridges is also involved in certain malformations of this type. However, the fact that a single agent can produce many different bulboventricular malformations by a single treatment at different points within the 48 h period between day 7 and 9 rather indicates that there is a fundamental link which is common to all the malformations observed.

Lev et al.[14] have presented a series of human hearts which exhibit many intermediate stages between normal development and complete transposition and they have included Fallot's tetralogy as one of these intermediates. Further study by Goor and Edwards[15] confirmed this and they enlarged this spectrum of anomalies by the inclusion of hearts exhibiting a double outlet right ventricle. Van Mierop and Wiglesworth have also investigated many hearts of these types and while they felt the spectrum should include the double-outlet right ventricle as an extreme stage, they would exclude transpositions altogether. However, individual litter analysis supports the concept of Lev et al.[14] as extended by Goor and Edwards[15].

No specimen appeared to exhibit all the features of the tetralogy because it is difficult to assess right ventricular hypertrophy. However, a number of hearts were observed which met all other necessary criteria including an overriding aorta and a pulmonary stenosis in the presence of mitral–aortic continuity.

Examples of this type of heart together with others with double-outlet right ventricle and classical complete transposition have all been seen within a single litter. Litters have also been found which contained all possible combinations of any two of the three major types of bulboventricular malformation. Each may also occur in isolation; a single litter may contain hearts which are either normal or exhibit only one of these abnormalities. Careful study of such a litter usually shows that there is a spectrum of abnormality within it; for example, the varied positions which the two main arteries may bear to each other in double-outlet right ventricle.

All three major types of defect have been seen after receiving retinoic acid on any of the four treatment days. However, two general trends may be observed; first, that hearts with a double-outlet right ventricle are most commonly seen after treatment on day 7 or 8; and, second, that overriding aortas are generally found after treatment on days $8\frac{1}{2}$, 9, or 10.

Although the timing of administration of the drug is of some significance, it is not crucial to the development of any one anomaly. This may presumably be due in part to the slightly different stages of development found in individual embryos in a litter at the time of dosage. It may also be a reflection of minor haemodynamic variations within each embryo, the position of the embryo within the uterus and the genetic heterogeneity of the hamsters used in this study. In addition, the ability of the heart to effect repairs during later stages of development should not be overlooked. Nevertheless, it is suggested that these findings are further evidence that the double-outlet right ventricle, classical transposition and 'overriding aorta complex' are not separate entities but rather represent parts along a spectrum of malformation.

Pexieder[5] recognizes that cell death is but one of several processes going on in the developing heart but believes that the *locus minoris resistentiae* resides in the bulbar and atrioventricular cushions. In particular, he suggests that the foci of death of undifferentiated mesenchyme found in these areas at certain stages of cardiogenesis have a morphogenetic role and that their intensity and topography normally depend on haemodynamic factors. Moreover, he suggests that those cells which are prenecrotic in the bulbar cushions may represent a common target for the action of biophysical as well as chemical teratogens. Such agents may increase or decrease the amount of death normally seen in any particular part of the heart.

In the present study greatly increased numbers of dying and dead mesenchymal cells have been observed in the bulbar cushions, and especially in their proximal portions, but a second outstanding ultrastructural feature of retinoic acid treated hearts is the large number of damaged and dead myocytes found in both the ventricle and the bulbus.

It would, therefore, seem reasonable to direct attention to both the muscular and cushion components of the developing bulboventricular area in considering possible mechanisms of teratogenesis.

Anderson et al.[16] have suggested that the group of bulboventricular malformations under consideration are all due to disordered conal development in association with changes in the position of the anterior ventricular septum. Assuming this to be true, the maldevelopment must be due to a change or defect in the amount or timing of one or more cellular processes such as proliferation or death. Similar alterations of migration of cells or the secretion of substances such as glycosaminoglycans must also be considered.

It is well established that vitamin A is required for normal fetal development and may be involved in fetal cell differentiation and organogenesis[18]. Ong and Chytil[19] have also suggested that retinoic acid is important in embryogenesis because retinoic acid binding protein levels fluctuate not only with cell type changes but also during morphogenesis. However the mechanism of action of vitamin A and related compounds is unknown at present. Hurley[20] has suggested that vitamin A may interfere with DNA synthesis and lengthen the cell cycle thus leading to a decreased rate of cell production.

Changes have been observed in the ultrastructure of a variety of cells and organs in culture and embryos *in vivo* following the administration of vitamin A or related compound. Loss of rough endoplasmic reticulum, mitochondrial swelling, accumulation of lipid droplets and the presence of autophagic vacuoles are common findings. Vacuolation of nuclear membranes, invagination and budding of the plasma membrane and cell death have also been prominent features of such reports[21]. Daniel et al.[22] have suggested that vitamin A in excess affects the integrity of cell membranes, and indeed increased lability of lysosomes[23] and permeability of mitochondria[24] in myocardial cells have all been demonstrated.

Kwasigroch and Kochhar[25] have described the low migration rate of vitamin A treated cells and increased adherance to substratum of epidermal cells following retinoic acid treatment has also been found[26].

In the latter report a greatly increased number of hemidesmosomes was observed and recently Elias and Friend[27] have found that retinoic acid is also able to stimulate the formation of both gap and tight junctions. This low migration rate may also be due to disturbance of those cells responsible for matrix formation[28]. In the heart tube and especially in the endocardial cushions this is important as it may interfere with the haemodynamics of the developing heart. This is very likely because retinoic acid has been implicated in the disordered biosynthesis of mucopolysaccharides and glycoproteins[29-31].

It has been noted earlier that myofibrillar development is slowed down by approximately 12 h after treatment with retinoic acid. This delay, when combined with the widespread death of myocardial cells, is quite likely to affect the functional capacity of the heart and this may itself be contributory to any defects produced. In addition, retinoic acid and related compounds are associated with abnormal vasculature in the limbs[32] and, if this is true of the pharyngeal arches also, this too may lead to the secondary development of cardiac abnormalities.

ACKNOWLEDGEMENTS

I wish to thank Mrs A. Agur, Dr M. J. Wiley and Mr C. G. T. Watterson for their help in this study, and Misses Pam Gale and Jill Parsons for secretarial assistance. This work was supported by the Medical Research Council of Canada and the Atkinson Charitable Foundation.

References

1. Fox, M. H. and Goss, C. M. (1956). Experimental production of a syndrome of congenital cardiovascular defects in rats. *Anat. Rec.*, 124, 189
2. Van Mierop, L. H. S. and Gessner, I. H. (1972). Pathogenetic mechanisms in congenital cardiovascular malformations. In: W. F. Friedman, M. Lesch and E. H. Sonnenblick (eds.). *Neonatal Heart Disease*, pp. 1–20. (New York and London: Grune and Stratton)
3. Patterson, D. F. (1968). Epidemiologic and genetic studies of congenital heart disease in the dog. *Circ. Res.*, 23, 171
4. Siller, W. G. (1958). Aortic dextroposition complexes in the fowl. A study in comparative pathology. *J. Pathol. Bacteriol.*, 94, 155
5. Pexieder, T. (1975). Cell death in the morphogenesis and teratogenesis of the heart. *Adv. Anat. Embryol. Cell Biol.*, 51, 3
6. Ruch, J. V., Zahnd, M., Robez-Kremer, G., Rumpler, O. and Gerlinger, P. (1965). Action tératogène plusieurs substances sur la morphogenèse cardiaque de l'embryon de poulet. *Arch. Biol.*, 76, 25
7. Khera, K. S. and Heggtveit, M. A. (1974). Fetal cardiovascular defects induced by thalidomide in the cat. *Teratology*, 9, 24
8. Kalter, H. and Warkany, J. (1961). Experimental production of congenital malformations in strains of inbred mice by maternal treatment with hypervitaminosis A. *Am. J. Pathol.*, 38, 1
9. Nolan, G. A. (1969). Variations in teratogenic response to hypervitaminosis A in three strains of the albino rat. *Food Cosmet. Toxicol.*, 7, 209
10. Robens, J. F. (1970). Teratogenic effects of hypervitaminosis A in the hamster and the guinea pig. *Toxicol. Appl. Pharmacol.*, 16, 88
11. Fantel, A. G., Shepard, T. H., Newell-Morris, L. L. and Moffett, B. C. (1977). Teratogenic effects of retinoic acid in pigtail monkeys. *Teratology*, 15, 65
12. Shenefelt, R. E. (1972). Morphogenesis of malformations in hamsters caused by retinoic acid: relation to dose and stage at treatment. *Teratology*, 5, 103
13. Taylor, I. M., Wiley, M. J. and Agur, A. (1979). Retinoic acid-induced bulboventricular malformations in the hamster. (In preparation)
14. Lev. M., Bharati, S., Meng, C. C. L., Liberthson, R. R., Paul, M. H. and Idriss, F. (1972). A concept of double outlet right ventricle. *J. Thoracic Cardiovasc. Surg.*, 64, 271
15. Goor, D. A. and Edwards, J. E. (1973). The spectrum of transposition of the great arteries: With specific reference to developmental anatomy of the conus. *Circulation*, 48, 406
16. Anderson, R. H., Arnold, R., Wilkinson, A., Becker, A. E. and Lubkiewicz, K. (1974). Morphogenesis of bulboventricular malformations. II. Observations on malformed hearts. *Br. Heart J.*, 36, 948
17. Van Mierop, L. H. S. and Wiglesworth, F. W. (1963). Pathogenesis of transposition complexes. II. Anomalies due to faulty transfer of the posterior great artery. III. True transposition of the great vessels. *Am. J. Cardiol.*, 12, 226
18. Ong, D. E., Page, D. L. and Chytil, F. (1975). Retinoic acid binding protein: occurrence in human tumors. *Science*, 190, 60
19. Ong, D. E. and Chytil, F. (1976). Changes in levels of cellular retinol- and retinoic-acid binding proteins of liver and lung during perinatal development of rat. *Proc. Natl. Acad. Sci. (USA)*, 73, 3976

20. Hurley, L. S. (1977). Nutritional deficiencies and excesses. In: J. G. Wilson and F. Clarke Fraser (eds.) *Handbook of Teratology*, Vol. 1, pp. 261–308 (New York and London: Plenum Press)
21. Morriss, G. M. (1973). The ultrastructural effects of excess maternal vitamin A on the primitive streak stage rat embryo. *J. Embryol. Exp. Morphol.*, **30**, 219
22. Daniel, M. R., Dingle, J. T., Glauert, A. M. and Lucy, J. A. (1966). The action of excess vitamin A alcohol on the fine structure of rat dermal fibroblasts. *J. Cell Biol.*, **30**, 465
23. Schweichel, J. U. and Merker, H. J. (1973). The morphology of various types of cell death in prenatal tissues. *Teratology*, **7**, 253
24. Wenzel, D. G. and Acosta, D. (1973). Permeability of lysosomes and mitochondria in cultured rat heart muscle and endotheloid cells as affected by vitamin A, chlorpromazine, amphotericin B and clofibrate. *Res. Comm. Chem. Pathol. Pharmacol.*, **6**, 689
25. Kwasigroch, T. E. and Kochhar, D. M. (1975). Locomotory behaviour of limb bud cells: effect of excess vitamin A in vivo and in vitro. *Exp. Cell Res.*, **95**, 269
26. Christophers, E. and Wolff, H. H. (1975). Differential formation of desmosomes and hemidesmosomes in epidermal cell cultures treated with retinoic acid. *Nature*, **256**, 209
27. Elias, P. M. and Friend, D. S. (1976). Vitamin A induced mucous metaplasia. An in-vitro system for modulating tight and gap junction differentiation. *J. Cell Biol.*, **68**, 173
28. Markwald, R. R., Fitzharris, T. P. and Adams Smith, W. N. (1975). Structural analysis of endocardial cytodifferentiation. *Dev. Biol.*, **42**, 160
29. Solursh, M. and Meier, S. (1973). The selective inhibition of mucopolysaccharide synthesis by vitamin A treatment of cultured chick embryo chondrocytes. *Calcif. Tissue Res.*, **13**, 131
30. Schimmelpfennig, K., Baumann, I. and Kaufmann, C. (1972). Studies on glycosaminoglycans (GAG) in mammalian embryonic tissue. II. Influence of vitamin A and Na-salicyate on embryonic GAG. *Naunyn-Schmiedeberg's Arch. Pharmacol.*, **272**, 65
31. DeLuca, L. and Wolf, G. (1970). Vitamin A and mucus secretion. A brief review of the effect of vitamin A on the biosynthesis of glycoproteins. *Int. Z. Vitamin Forsch.*, **40**, 284
32. Fraser, B. A. (1977). The relationship of aberrant vasculogenesis to retinoic acid induced dysmelia in the hamster fetus. *Presented at the 5th International Conference on Birth Defects*, August 21–27, Montreal

8
Physiological cell death in normal and abnormal rodent limb development

W. J. SCOTT, JR.

INTRODUCTION

It seems well established that many teratogenic agents induce malformations as a result of their cytotoxic properties (reviewed by Scott)[1]. Abnormal loss of cells due to necrosis can easily be envisaged as a hindrance to normal development. Interestingly, it also appears that abnormal development may result when cells which normally die during development are prevented from doing so at the proper time.

The purpose of this presentation is to consider three episodes of physiological necrosis which occur during normal limb development in rodents and describe alterations in the quantity and timing of these episodes in cases of abnormal limb development.

APICAL ECTODERMAL RIDGE

The specialized thickening of ectoderm capping the limb bud distally is known as the apical ectodermal ridge or AER. This 'organ' is thought by most investigators to possess inductive properties which permit the underlying mesoderm to form normal limb structures.

For some time it has been clear that mouse[2,3] and rat[3-5] AER contain necrotic cells. Milaire[3] has presented diagrammatic representation of this phenomenon in mouse and rat fore- and hindlimbs. From that diagram one gains the idea that cell death within the AER has a precise spatiotemporal distribution. At first two separate and distinct zones are seen, a diffuse postaxial zone and a more discrete preaxial zone. With the passage of time the preaxial AER necrosis expands. In our rat strain[5] these zones finally coalesce so that a continuous band of necrosis surrounds the limb tip (Figure 8.1) corresponding more closely to the pattern depicted for the mouse by Milaire.

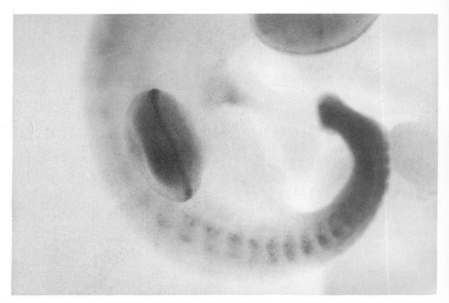

Figure 8.1 Uptake of Nile blue indicating dead cells in the hindlimb apical ectodermal ridge of a rat embryo at 9 a.m. of day 13. Note the continuous band of staining surrounding the apex of the limb

The most important question regarding physiological cell death in the AER is the role of this necrotic episode in normal limb development. Necrosis within the AER has most often been equated with involution of this transient organ even though dead cells are evident long before the AER has attained its peak size[2]. One might speculate that the loss of cells by degeneration acts to decrease the inductive activity of the AER on the underlying mesoderm. Since the exact mechanism of AER induction is unknown it is impossible to determine directly whether necrosis within this structure decreases its inductive activity. However, some indirect evidence in favour of this hypothesis does exist. In one case of inherited polydactyly[6,7] and in numerous cases of drug-induced preaxial polydactyly[5,8,9], the episode of preaxial AER necrosis is delayed in onset while the AER retains a thickened, 'active' appearance. This has led to the suggestion that this lack of cell death leads to a stronger inductive effect of the preaxial AER with the formation of an extra digit as the result[5]. Further evidence for this suggestion is found in the work of Rooze[7] who demonstrated an increased content in basal RNA and alkaline phosphatase activity in the preaxial AER of dominant hemimelia heterozygote (Dh/+) mouse limbs, contrary to what happens in normal limbs. The mutant limbs are devoid of physiological necrosis and many will later develop a triphalangic hallux or frank preaxial polydactyly. At odds with this idea are the results of high gene[7] or drug[8] dosage which lead to absence of preaxial digits. In these cases preaxial AER cell death is likewise delayed in onset.

It is also profitable to inquire into the basis of physiological cell death within the AER. Milaire[10] and Rooze[7] imply that apical ectodermal maintenance factor[11] produced by the limb mesoderm keeps the AER cells viable. The appearance of physiological necrosis would thus represent a lowering of AEMF concentration in the subjacent mesoderm.

Results from our drug-induced polydactyly studies have led us to suggest another possibility, namely, that the mesodermal cells which are killed by these agents normally produce a substance lethal to the AER cells. In the absence of the mesodermal cells, the ectodermal cells continue to live and to influence digital development. Evidence for this speculative idea was gained from study of four cytotoxic agents which induce a high percentage of polydactyly in rat fetuses. The optimal time of agent administration for induction of polydactyly varied between all four agents as did the rapidity of cytotoxic activity of each agent as indicated in Table 8.1. When these numbers are

Table 8.1 Polydactyly-inducing and cytotoxic properties of four drugs studied in rat embryos

Drug	Optimal time of administration to induce polydactyly (h)	Interval between drug administration and appearance of moderate cytotoxic response (h)	Sum
6-Mercaptopurine	264	24	288
Cytosine arabinoside	276	15	291
Cyclophosphamide	280	11	291
5-Fluorodeoxyuridine	283	9	292

added, their total coincides closely with the expected time of onset of physiological cell death in the ectoderm of control embryos. For example, cytosine arabinoside is most effective in producing polydactyly when administered at 276 h of rat development (Table 8.1[12]). It takes about 15 h for a moderate cytotoxic response to become evident. The sum of 276 + 15 or 291 h of development coincides with the time of onset of physiological necrosis in the hindlimb ectoderm. Thus we believe that the mesodermal cells must be inactivated (killed) before they have produced or delivered a sufficient amount of the postulated lethal factor to the AER. Using similar information, Milaire (personal communication) has suggested that necrosis of the mesodermal cells could lead to exaggerated liberation of AEMF which then prevents physiological necrosis in the preaxial ectoderm.

Another explanation of this data depends on information on finite lifespan of cells in culture[13]. In this case the cells degenerate after a certain number of mitotic divisions. If a similar situation holds in the cells of the AER, then the delay in onset of necrosis could be related to the biochemical effect of these agents, all of which inhibit DNA synthesis and thus delay mitotic division. When these cells recover and pass through the required mitotic divisions they would then die somewhat later in development.

Insufficient information is presently available to choose among these three possibilities.

FOYER PRIMAIRE PREAXIAL

Milaire[14] was the first to describe a zone of physiological necrosis appearing in the deep preaxial mesoderm and he termed it the *foyer primaire preaxial* or 'fpp'[15]. This zone of necrosis has been found in the fore- and hindlimbs of rat and mouse embryos, the only species in which it has been searched for besides the mole. This zone of necrosis has a very specific location at the most proximal extent of the apical ectodermal ridge just medial to the marginal venous sinus in the preaxial mesoderm (Figure 8.2).

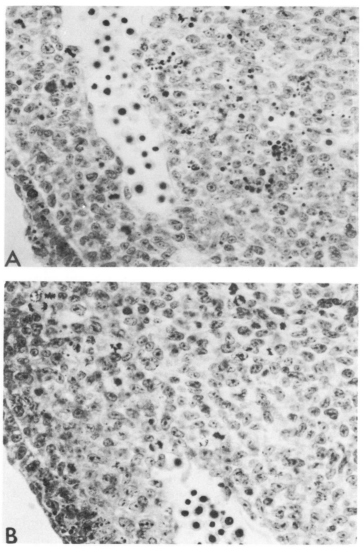

Figure 8.2 Tangential section through hindlimb at 324 h of development. ×423. A, Control embryo. Note large number of degenerating cells in preaxial mesoderm beneath marginal venous sinus. Also note location at the most proximal extension of preaxial AER. B, Cytosine arabinoside treated embryo. Note absence of degenerating cells in the same area. (Reprinted from Scott et al.[5])

The function of the 'fpp' in limb development seems quite clear, due in large part to the work of Milaire[6, 15] and his colleague Rooze[7]. This zone of necrosis acts to reduce the size of digit 1 by way of decreasing the cellular contribution of mesoderm to this digit. In three different cases of excess preaxial digits this zone is absent or reduced in size.

(1) Heterozygote dominant hemimelia (Dh/+) mouse embryos often have a triphalangic hallux or less frequently preaxial polydactyly[7, 16]. The 'fpp' in these embryos is also missing and presumably these cells which do not die as in normal embryos contribute to the excess digital material.

(2) Moles normally develop an extra preaxial digit on the fore- and hindlimbs (prepollex and prehallux) and the 'fpp' is again missing[15].

(3) It has been shown that a polydactyly-inducing regimen of cytosine arabinoside leads to an absence of the 'fpp' in rat embryos[5] and presumably the lack of normal cell death in the preaxial mesoderm leads to preaxial digital excess.

The converse situation, that is increased size of the 'fpp' with a resultant loss of preaxial digits, has been shown to occur in heterozygotes of the mouse mutant oligosyndactylism[6, 16, 17]. Strangely it is digits 2 and 3 which are most often fused or missing. Digit 1 is usually normal or in a peculiar variant there is duplication of the hallux combined with absence of digit 2[16].

Interfering with complete acceptance for this role of the 'fpp' in normal and abnormal limb development are the results of higher gene or drug dosage. Homozygote dominant hemimelia mouse embryos (Dh/Dh) most often have reduced numbers of preaxial digits as do some heterozygotes[16]. In Dh/+ embyros with deficient limbs Rooze[7] found the preaxial mesoderm devoid of necrotic cells. Likewise, higher drug dosage[8] can lead to preaxial ectrodactyly yet the 'fpp' is absent as it is with lower polydactyly inducing doses. Also discomforting is the reduction or absence of the forelimb 'fpp' in Dh/+ embryos which nevertheless develop quite normally[7]. These contradictory situations need a logical explanation before the true role of this necrotic episode in development is fully understood.

Speculation about the cause of death in the cells of the 'fpp' rests mainly on its location. As mentioned earlier this zone of necrosis lies beneath the most proximal extremity of the preaxial AER. Milaire[15] has used this information to suggest that the waning influence of the AER leads to the death of these cells. This supposition is supported by the drug-induced polydactylies where the preaxial AER is enlarged and the 'fpp' does not appear.

The idea of apical ectodermal ridge producing a substance which prevents degeneration of mesodermal cells has strong experimental support especially from the work of Cairns[18]. Cell death was obvious in blocks of chick limb mesoderm cultured alone or covered by non-AER limb ectoderm. On the other hand, when similar blocks were covered with AER or even cultured some distance away, cell deaths did not occur.

Recently we have uncovered another situation in which the 'fpp' cells do not die but in contrast to other such cases there is no observable effect on the AER. The thymidine analogue, BUdR, given to pregnant rats on day 12 of gestation produces a high incidence of preaxial polydactyly[19]. Examination of limb buds from embryos treated with this agent revealed absence of the 'fpp', but cell death patterns in the AER were unchanged from the control pattern. This suggests that the onset of death in cells of the 'fpp' can be altered intrinsically and is not under total control of the AER.

INTERDIGITAL NECROTIC ZONE

Between the developing digital rays of the fore- and hindfoot in mouse embryos two separate episodes of necrosis originate, one proximal and one distal[17]. These zones spread and finally coalesce to form the triangular interdigital necrotic zone. Simultaneously the marginal ectodermal overlying the interdigital zones shows degenerative changes while that overlying the digits retains a healthy 'active' appearance[17]. This sequence of events appears much the same in chick embryos[20]. However, Cameron and Fallon[21] have shown that interdigital necrosis is not present during amphibian limb development.

The function of this necrotic zone seems quite clear, that is, to remove the interdigital tissue thus allowing digits to become free and act independently. This idea is strengthened by the fact that this episode is absent in species with webbed feet such as the duck[22,23].

In terms of teratogenesis, then, this zone of necrosis would seem important in conditions leading to syndactyly. Certainly this is the case with Janus green induced syndactyly in chick embryos[20,23,24]. In experimental mammalian teratology the interdigital necrotic zone has not yet been implicated as the primary cause of syndactyly. Milaire has thoroughly studied two cases of genetic syndactyly[17] and one case of drug-induced syndactyly[25]. Although interdigital necrosis did not follow the normal pattern, he found earlier developmental alterations to be the basis of fusion. This is not surprising since the syndactylous conditions observed by Milaire were mainly primary fusions involving skeletal elements rather than soft tissue syndactyly. Many chemical agents are capable of inducing soft tissue syndactyly but none has yet been studied in the detailed manner necessary to determine the primary cause. Surely some of these agents will act like Janus green in the chick embryo.

The molecular mechanism(s) responsible for the death of interdigital mesodermal cells is unknown. Again, however, some thought needs to be given to the relation between mesoderm and ectoderm. Milaire[17] has shown that the high RNA content and many enzymatic activities of the AER suddenly decrease in the interdigital ectoderm immediately prior to mesodermal necrosis. The author suggests that this may be due to a sudden decrease in maintenance activity of the mesoderm. Subsequently the marginal mesodermal cells in the interdigital area show signs of degeneration followed closely by similar cytolytic changes in the overlying ectoderm. The death of these cells in the AER may be due to the underlying mesodermal cell death as suggested by Kieny et al.[20] in chick embryos.

Other episodes of physiological necrosis occur during rodent limb development but will not be discussed here due to limited knowledge about their function in normal development or a lack of understanding of their involvement in limb malformations.

In summary, three episodes of physiological necrosis which occur in the developing rodent limb have been discussed. Each plays an integral role in normal development and an upset in the schedule of these episodes can lead to limb malformations.

The molecular mechanisms responsible for death of cells within these zones are unknown. The small amount of evidence available suggests that tissue interactions between ectoderm and mesoderm are important determinants of whether these cells live or die. Future work will need to uncover the exact nature of these interactions in order to advance our understanding of these enigmatic phenomena.

ACKNOWLEDGEMENTS

This work was supported by NIH grant HD06526.

References

1. Scott, W. J. (1977). Cell death and reduced proliferative rate. In: J. G. Wilson and F. C. Fraser (eds.). *Handbook of Teratology*, Vol. 2, pp. 81–98. (New York: Plenum Publishing Corp.)
2. Jurand, A. (1965). Ultrastructural aspects of early development of the fore-limb buds in the chick and the mouse. *Proc. R. Soc. London*, 162, 387
3. Milaire, J. (1967). Evolution des processuses dégénératifs dans la cape apicale au cours du développement des membres chez le rat et la souris. *C. R. Acad. Sci.*, 265D, 137
4. Schweichel, J. U. (1972). Das electronenmikroskopische Bild des Abbaues der epithelialen Scheitelleiste während der Extremitätenentwicklung bei Rattenfeten. *Z. Anat. Entwickl.*, 136, 192
5. Scott, W. J., Ritter, E. J. and Wilson, J. G. (1977). Delayed appearance of ectodermal cell death as a mechanism of polydactyly induction. *J. Embryol. Exp. Morphol.*, 42, 93
6. Milaire, J. (1971). Evolution et déterminisme des dégénérescences cellulaires au cours de la morphogenèse des membres et leurs modifications dans diverses situations tératologiques. In: H. Tuchmann-Duplessis (ed.). *Malformations Congénitales des Mammifères*, pp. 131–149. (Paris: Masson et Cie)
7. Rooze, M. A. (1977). The effects of the Dh gene on limb morphogenesis in the mouse. In: D. Bergsma and W. Lenz (eds.). *Morphogenesis and Malformation of the Limb*, pp. 69–95. (New York: Alan R. Liss, Inc.)
8. Scott, W. J., Ritter, E. J., Wilson, J. G. and Schreiner, C. (1977). Pathogenesis of 6-mercaptopurine (6-MP) induced digital defects. *Excerpta Medica*, 426, 60
9. Klein, K. L., Scott, W. J. and Wilson, J. G. (1978). Aspirin induced teratology: a distinctive pattern of preaxial cell death and polydactyly in the rat. *Teratology*, 17, 45A
10. Milaire, J. (1977). Histochemical expression of morphogenetic gradients during limb morphogenesis. In: D. Bergsma and W. Lenz (eds.). *Morphogenesis and Malformation of the Limb*, pp. 37–67. (New York: Alan R. Liss, Inc.)
11. Zwilling, E. and Hansborough, L. A. (1956). Interaction between limb bud ectoderm and mesoderm in the chick embryo. III. Experiments with polydactylous limbs. *J. Exp. Zool.*, 132, 219
12. Scott, W. J., Ritter, E. J. and Wilson, J. G. (1975). Studies on induction of polydactyly in rats with cytosine arabinoside. *Dev. Biol.*, 45, 103
13. Hayflick, L. (1965). The limited in vitro lifetime of human diploid cell strains. *Exp. Cell Res.*, 37, 614
14. Milaire, J. (1963). Etude morphologique et cytochimique du développement des membres chez la souris et chez la taupe. *Arch. Biol. (Liege)*, 74, 129
15. Milaire, J. (1976). Rudimentation digitale au cours du développement normal de l'autopode chez les mammifères. In: A. Raynaud (ed.). *Mécanismes de la Rudimentation des Organes chez les Embryons de Vertebres*. (Paris: Editions du CNRS)
16. Grüneberg, H. (1963). *The Pathology of Development*. (New York: John Wiley and Sons, Inc.)

17. Milaire, J. (1967). Histochemical observations on the developing foot of normal, oligosyndactylous (Os/+) and syndactylous (sm/sm) mouse embryos. *Arch. Biol.*, 78, 223
18. Cairns, J. M. (1975). The function of the ectodermal apical ridge and distinctive characteristics of adjacent distal mesoderm in the avian wing-bud. *J. Embryol. Exp. Morphol.*, 34, 155
19. Scott, W., Ritter, E., Wilson, J. and Schreiner, C. (1978). A second mechanism of polydactyly induction. *Teratology*, 17, 37A
20. Kieny, M., Pautou, M.-P. and Sengel, P. (1976). Limb morphogenesis as studied by Janus green B- and vinblastine-induced malformations. In: M. Marois and J. Ebert (eds.). *Tests of Teratogenicity* In Vitro, pp. 389–415. (Amsterdam: North Holland)
21. Cameron, J. and Fallon, J. F. (1977). The absence of cell death during development of free digits in amphibians. *Dev. Biol.*, 55, 331
22. Deleanu, M. (1965). Toxic action upon physiological necrosis and macrophage reaction in the chick embryo leg. *Rev. Roumaine Embryol. Cytol.*, 2, 45
23. Saunders, J. W. and Fallon, J. F. (1966). Cell death in morphogenesis. In: M. Locke (ed.). *Major Problems in Developmental Biology*, pp. 289–314. (New York: Academic Press)
24. Menkes, B. and Deleanu, M. (1964). Leg differentiation and experimental syndactyly in chick embryo. *Rev. Roumaine Embryol. Cytol.*, 1, 69
25. Milaire, J. (1971). Étude morphogénétique de la syndactylie postaxiale provoquée chez le rat par l'hadacidine. II. Les bourgeons de membres chez les embryons de 12 à 14 jours. *Arch. Biol.*, 82, 253

9
Palate morphogenesis: role of contractile proteins and neurotransmitters

E. F. ZIMMERMAN

INTRODUCTION

Environmental agents, such as drugs and pollutants, can cause congenital malformations when the fetus is exposed to them at a critical period during development at an optimal fetal concentration. Although much research has been carried out to elucidate the teratogenic mechanisms, little knowledge exists on the precise mechanisms involved. Thus, although a teratogen may be shown to inhibit or affect a biochemical pathway in a fetal tissue, usually it is not known how this biochemical lesion expresses itself in the observed malformation. In some instances the teratogen affects morphogenesis. Glucocorticoids have been observed to delay palate shelf rotation[1] and causes myopathy in adult animals and humans[2]. Therefore, it would seem possible that glucocorticoids exert their teratogenic effects on a contractile system. However, although an 'intrinsic shelf force' within the palate has been postulated to move the shelf[3], the actual mechanism of palate shelf rotation has not been elucidated. Therefore, for the last few years we have been studying the mechanisms by which palate shelves undergo their morphogenetic movement in order to be able to study mechanisms of teratogenic action. This research has explored the possibility that the 'intrinsic shelf force' is derived from the function of contractile proteins organized into muscle or non-muscle systems.

RESULTS
Biochemistry of actin and myosin

The presence and rate of synthesis of actin and myosin in the palate of the A/J mouse, just prior to shelf rotation (day 14.5), was determined. It was shown that actin and myosin were synthesized in the palate at a rate equal to that of the developing skeletal muscle tissue, the tongue[4]. In retrospect, it was not surprising that palatal cells contained the contractile proteins; all cells do[5].

However, the significant rate of synthesis suggested that contractile systems in the palate, muscle or non-muscle, might play a role in shelf elevation.

Morphological analysis of the vertical palate

Myosin ATPase histochemistry was performed on frozen sections of day 14.5 fetal mouse heads. Three areas of the palatal mesenchyme gave a positive reaction: (1) a reaction product typical of skeletal muscle on the oral side of the posterior palate (region 1); (2) a 'heavy-diffuse' reaction product on the tongue side extending from the top mid-palate to the posterior limit (region 2) (Figure 9.1a); and (3) a 'light-diffuse' reaction product along the oral epithelium present in mid-palate extending towards the anterior limit (region 3)

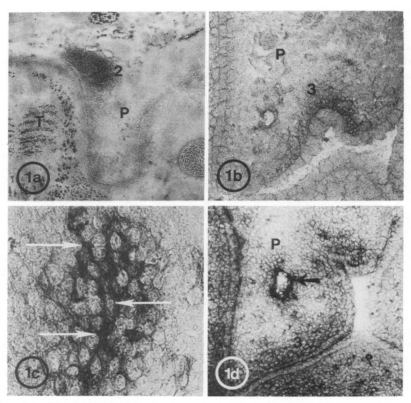

Figure 9.1 Myosin ATPase histochemistry of day 14.5 fetal heads. Reaction was carried out by method of Padykula and Herman[7] for 30 min unless otherwise indicated. (a) A cryosection of the posterior end of the palate (P) showing a positive reaction product in region 2. T, tongue. ×49. (b) An anterior palate shelf with a positive reaction in region 3. ×98. (c) In another experiment using 5 mM BAL, the myosin ATPase reaction product in region 2 is deposited primarily in the cytoplasmic processes of the cells (arrows). ×385. (d) The myosin ATPase reaction product in region 3 and the epithelium in the presence of 5 mM BAL and incubated for 90 min. A positive reaction is also associated with a palatine vessel (arrow). ×112. The longer incubation time is responsible for the strong reaction

(Figure 9.1b)[6]. However, the specificity of the histochemical reaction for myosin ATPase has been questioned, since other phosphatases may act on the ATP substrate[7]. Therefore, the histochemical reaction was repeated in the presence of 5 mM BAL, an inhibitor of alkaline phosphatase. The reaction was still positive in regions 2 (Figure 9.1c) and 3 (Figure 9.1d) and the reaction product could be observed in structures coursing between the cells (Figure 9.1c). To obtain further support for the presence of contractile proteins in cells of regions 2 and 3, cryostat sections were incubated with antibodies to smooth muscle myosin and skeletal muscle actin and indirect immunofluorescence studies carried out. Mesenchymal cells in regions 2 and 3 and the epithelium showed increased fluorescence with anti-actin compared to other areas of the palate (Figure 9.2b)[8,9]. Fluorescence with anti-myosin was observed in the filopodial processes of the region 2 and 3 cells (Figures 9.2c, d). It is interesting to note that nerves course through regions 2 and 3 (Figures 9.2a, b) which could possibly exert a trophic influence on the development of these systems. Krawozyk and Gillon[10] also showed an enhanced fluorescence reaction in the epithelium and some in the mesenchyme using a human anti-smooth muscle serum.

Light and electron microscopic studies confirmed the presence of a mass of developing skeletal muscle in region 1[6]. In the area of region 2, a distinct cellular condensation was seen in light micrographs (Figure 9.2a). This area consists of a small centre of ossification surrounded by an unorganized web of undifferentiated mesenchymal cells. These mesenchymal cells extend upwards toward the nasal septum, abutting the perichondrium of the nasal septum and cranial base cartilage, as well as extending into the shelf toward the medial edge. Ultrastructurally, these cells possess filopodia-like projections (Figure 9.3a). Bundles of 70 Å microfilaments were observed just beneath the plasma membrane (Figure 9.3b) and packed in the filopodia-like projections (Figure 9.3c). The cells and projections are connected by specialized junctions, including zonula adherens, forming a putative cohesive contractile network. In region 3, a striking orientation was observed of the putative contractile cells aligned perpendicular to the oral epithelium and extending one-third into the shelf (Figure 9.3d). These cells contain similar filamentous systems as seen in region 2 by electron microscopy. Projections from region 3 cells were found embedded in the basement membrane material of the oral epithelium. The presence of the large amounts of 70 Å microfilaments in the cells and their filopodia in regions 2 and 3 mesenchyme are in accord with the observations from the myosin ATPase, and myosin and actin immunofluorescence experiments. Axonal bundles and single axons were commonly observed coursing through regions 2 (not shown) and 3 (Figure 9.3d), often in close association with mesenchymal cells. Taken together, these results suggest that the contractile proteins in regions 2 and 3 in the palate are organized as primitive non-muscle contractile systems. Figure 9.4 shows a three-dimensional view of these systems. Innes[11] has confirmed the presence of skeletal muscle and cells with many filopodial processes in the palate mesenchyme before palate shelf rotation.

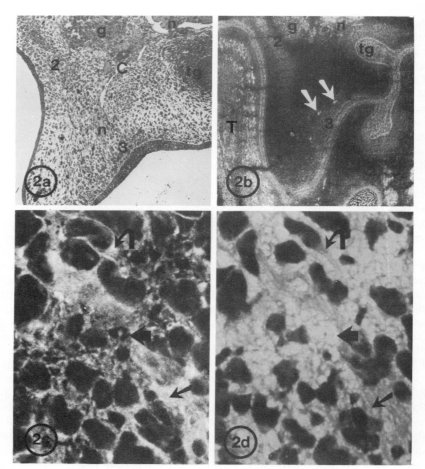

Figure 9.2 Localization of contractile proteins in day 14.5 palate by immunofluorescence. a, A light micrograph of a 1 μm section of an EM blockface, just anterior to mid-palate. The anterior limit of region 2 is seen along the top tongue side epithelium. The mesenchymal cells along the oral epithelium, region 3, show a distinct orientation radiating out from the epithelium toward the middle of the palate. C, capillary; g, ganglion; n, nerves; tg, tooth germ. ×91. b, Anti-actin immunofluorescence of day 14.5 fetal heads. Cryosections were incubated with an IgG preparation of rabbit anti-mouse skeletal muscle actin. After washing, slides were incubated with fluorescein-tagged goat anti-rabbit IgG. Fluorescence at mid-palate. g, ganglion; n and arrows, nerves; T, tongue. tg, tooth germ. ×70. c, Immunofluorescence of region 3 in palate employing anti-chicken gizzard myosin IgG. ×945. d, Same section as (c) restained with Richardson's stain. Note areas of bright fluorescence corresponding to areas with high concentrations of cellular projections (straight and curved arrows). An area of diminished fluorescence corresponds to a region with few cellular projections (short arrows). ×945

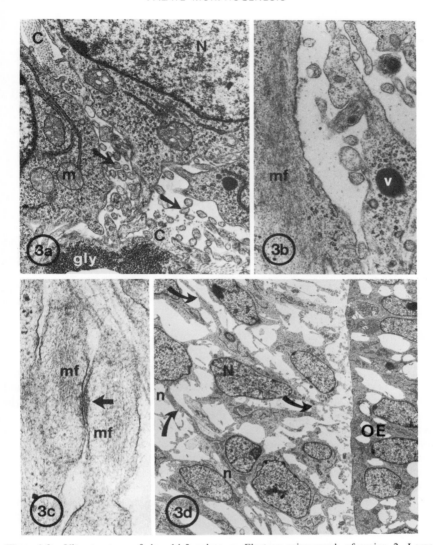

Figure 9.3 Ultrastructure of day 14.5 palate. a, Electron micrograph of region 2. Large numbers of filopodial processes (arrows) and collagen bundles (c) that are coursing between the cells cut in cross-section; gly, glycogen; m, mitochondrion; N, nucleus. × 12 040. b, An electron micrograph of the mesenchymal cells of region 2. Broad bands of 70 Å microfilaments (mf) are seen in the area of the plasma membrane of these cells. Also observed are numerous dense core vesicles (v). × 22 820. c, Filopodial-like projections seen in region 2 packed with 70 Å microfilaments (mf) and connected by an adherens junction (arrow). Microfilaments are also seen along the plasma membrane of an adjacent cell. × 47 600. d, A survey electron micrograph of region 3. The mesenchymal cells radiate out from the oral epithelium and possess elongated cellular projections (arrows) which form a network of cells and projections. n, nerves; N, nucleus. × 3010

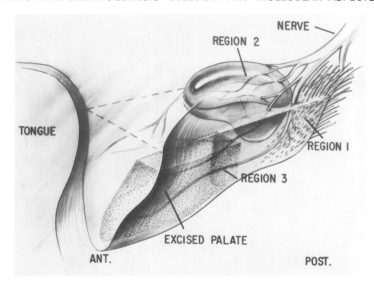

Figure 9.4 Schematic drawing of the contractile systems in the day 14.5 vertical palate. While nerves are shown coursing through regions 1, 2 and 3, this is not a precise representation of their anatomical location

Morphology of palate shelves during rotation

Cells of regions 2 and 3 during rotation were analysed for changes in cell shape and their orientation to the epithelium. Embryos were collected at day 15.5, and separated into those whose palates had rotated but not fused and those whose palates had rotated and fused. Palates were analysed by light microscopy in plastic sections and also by electron microscopy. Since collections of embryos at day 14.5 and day 15.5 only rarely contained embryos in the process of elevation, experiments were also carried out using embryos cultured *in vitro*[12]. Employing this whole embryo culture system, palates in various states of rotation could be consistently obtained. In these experiments, the anterior palate had rotated about halfway to the horizontal position (palate shelf index (PSI = 3.0) after 2 h and three-quarters of the way (PSI = 4.0) after 4 h). Posteriorly, an average PSI of 2.0 after 2 h and 3.0 after 4 h were obtained.

Characteristic changes in the shape of the palate and changes in the shape of the mesenchymal cells and their orientation to the epithelium are summarized (Figure 9.5). In region 3, before palate rotation, cells have an elongated shape and are aligned perpendicularly to the oral epithelium. After the area of the palate moved, the cells rounded and lost their perpendicular orientation. Although the tongue side opposite and tongue side bend cells (region 2 and pre-region 2) were round before rotation, they have assumed an elongate shape and are oriented perpendicularly to the epithelium after the oral side of the palate rotated. Thereafter, during rotation, the tongue side epithelium pulled inwards and the mesenchymal cells assumed a round shape, losing their orientation to the epithelium. It should be noted that comparable cell shape

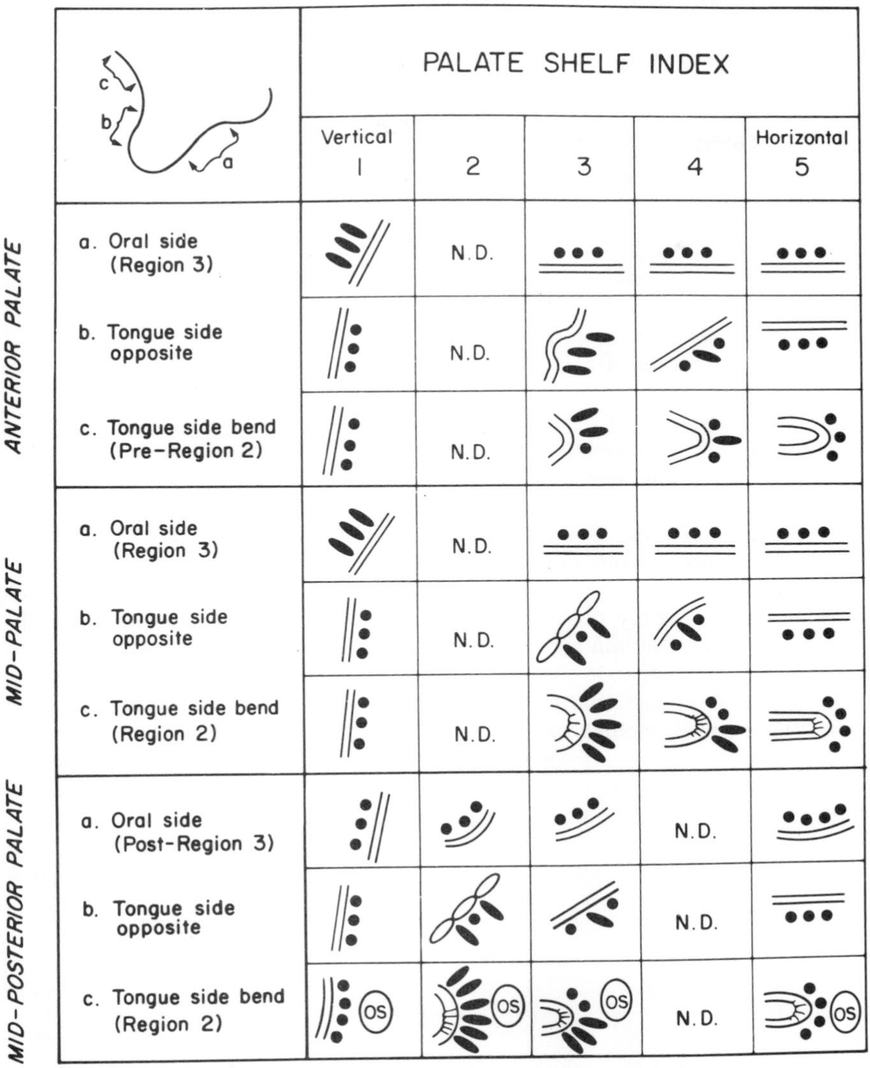

PALATE MORPHOGENESIS

Figure 9.5 A summary of the changes in cell shape and orientation that occurred in mesenchymal and epithelial cells of three specific sections during palate rotation. The cross-sectional views of the palate at the anterior, mid- and mid-posterior ends are indicated. The three areas are: (a) the oral side (region 3 or post-region 3); (b) the tongue side opposite to the oral side; and (c) the tongue side at the bend (pre-region 2 and region 2). Diagrams of the position of the epithelium (==) and the shape and orientation of the mesenchymal cells (filled symbols) to the epithelium are drawn at the various stages of rotation, using the palate shelf index as the quantitative measure of rotation[12]. Cellular shape and orientation of cells at PSI of 1 are derived from day 14.5 vertical palates (*in vivo*) and PSI of 5 from day 15.5 fused palates (*in vivo*). N.D., not determined

changes did not occur in the interior mesenchyme. These observations suggest that contraction of peripheral mesenchymal cells, including those of regions 2 and 3 and possibly the epithelium, were supplying the 'intrinsic shelf force' to rotate the palate.

ATP-induced palate shelf rotation in glycerinated mouse heads and inhibition by cytochalasin B

Glycerinated muscle and some non-muscle cells have been demonstrated by other workers to contract by the addition of ATP[13,14]. To obtain further evidence that contractile proteins are involved in palate morphogenesis, day 14.75 heads were glycerinated at $-20\,°C$ in glycerol–standard salt solution for three weeks or more and the effect of ATP was tested after deglycerination and tongue excision. A solution of 5 mM ATP stimulated palate shelf rotation optimally after a 30 min incubation at $25\,°C$: anterior, $p < 10^{-6}$, posterior, $p < 0.05$ (Table 9.1); rotation decreased thereafter. A dose–response experiment with ATP showed that anterior rotation was stimulated maximally at 5 mM ATP; stimulation decreased at higher concentrations. Posterior rotation was maximal at 10 mM ATP. Whether differences between the anterior and posterior shelves are significant is not known. Light microscopy of the stained paraffin sections indicated that ATP stimulated the contraction of the cytoplasmic processes in the mesenchyme. Also, all the mesenchymal cells now could be readily observed to be attached to one another, as in a syncytium[15]. When the number of contracted cytoplasmic processes was plotted against palate shelf index, it was seen that the amount of cytoplasmic contraction was proportional to palate rotation. Furthermore, the slopes of the lines for the oral, medial and tongue side cells in the peripheral mesenchyme were equal and greater than the slopes of those cells in the internal mesenchyme. These results suggest that contraction of the cytoplasmic processes in the peripheral mesenchyme was responsible for palate morphogenesis.

Table 9.1 Stimulation of palate shelf rotation by ATP and reversal with cytochalasin B in glycerinated mouse heads*

		Palate shelf index	
Condition	n	Anterior	Posterior
Control – zero time	24	2.00 ± 0.12	1.12 ± 0.07
Control – incubation	16	2.06 ± 0.14	1.25 ± 0.11
5 mM ATP	16	3.81 ± 0.19	1.69 ± 0.18
Cytochalasin B (2 μg/ml) + 5 mM ATP	24	2.88 ± 0.19 ($p < 0.002$)	1.79 ± 0.15
Cytochalasin B (20 μg/ml) + 5 mM	20	2.05 ± 0.15 ($p < 10^{-6}$)†	1.35 ± 0.13

* Day 14.75 embryo heads were glycerinated in 50% glycerol–standard salt solution (0.1 M KCl, 1mM MgCl$_2$, 10 mM potassium phosphate, pH 7.0) for three weeks or more at $-20\,°C$. After deglycerination of heads, tongues and mandibles were excised, and zero time controls were immediately fixed. All other heads were preincubated with standard salt solution containing 0.2% ethanol or cytochalasin B in ethanol at $25\,°C$ for 10 min. ATP or equivalent amount of H$_2$O (0.5%) was added and heads were incubated for 30 min at $25\,°C$. Values are means $\pm$ standard errors
† p values obtained by comparison with ATP + 0.2% ethanol by Student's t-test

Cytochalasin B has been reported to inhibit a wide variety of cellular movements mediated by microfilaments. Therefore, the glycerinated heads were pretreated with cytochalasin B (20 μg/ml) and the rotation of the ATP-induced palate measured. Cytochalasin B pretreatment inhibited rotation, predominantly at the anterior shelf: anterior, $p < 10^{-6}$, posterior, $p < 0.15$ (Table 9.1). Lower concentrations of cytochalasin B produced less inhibition of ATP-induced palate rotation.

Cytoplasmic extracts containing actin show temperature and nucleotide specificity in gelation and contraction[16,17]. When the ATP-induced reaction in glycerinated heads at low temperature (6 °C) was determined, palate shelf rotation was markedly inhibited. Similarly, palate rotation showed nucleotide specificity: GTP was less effective than ATP, and ADP and CTP were ineffective.

In summary, these results suggest that contraction of the cytoskeleton composed of microfilaments in peripheral mesenchymal cells plays a major role in palate morphogenesis.

Palate rotation in embryo culture and pharmacological studies

To test whether skeletal, smooth or non-muscle contractile systems function during palate movement, effects of agonists and antagonists to these contractile systems were measured. First, an embryo culture system was developed to facilitate these studies. Mouse embryos with tongues removed (day 14.75) were cultured close to the time that palatal shelves move *in vivo*. The anterior end of the palate shelf completely rotated after overnight culture. However, rotation of the posterior end of the palate was only partial. Table 9.2 summarizes the effects of neurotransmitters and their antagonists on palate shelf rotation in embryo culture. In a time-course study, it was observed that palate rotation in the culture system was completed by 6 h. Cholinergic agents,

Table 9.2 Effect of agonists and antagonists of neurotransmitters on palate shelf rotation in embryo culture*

		Palate shelf rotation	
Agonist	Antagonist	Anterior	Posterior
Serotonin		Increase	
	Methysergide	Decrease	
Bethanechol			Increase
Pyridostigmine			Increase
Carbachol			Increase
	Atropine†		
	Curare		Increase
	Hexamethonium		Decrease

* Embryos were incubated[12] for 2 h in the presence of serotonin (10^{-5} M), methysergide (10^{-4} M), bethanechol (10^{-8} M), pyridostigmine (10^{-5} M), carbachol (10^{-5} M), and hexamethonium (10^{-6} M) and incubated overnight with bethanechol (10^{-8} M), carbachol (10^{-5} M), hexamethonium (10^{-4} M), and curare ($2 + 10^{-3}$ M) to produce a significant effect on palate rotation at least at the $p < 0.05$ level
† Embryos were incubated with atropine at 10^{-4} M for 2 h and 3×10^{-5} M drug overnight without a significant effect on palate rotation

pyridostigmine ($2 \times 10^{-6} - 9 \times 10^{-5}$ M) and bethanechol ($10^{-10} - 10^{-4}$ M), both enhanced posterior shelf rotation after overnight culture. Pyridostigmine (9×10^{-5} M) increased posterior shelf rotation 74% over control, bethanechol (10^{-4} M) 53%. Bethanechol stimulated posterior shelf by 60% within 60 min[12]. Carbachol also stimulated posterior shelf rotation. These results implied that the neurotransmitter, acetylcholine, could exert a control over posterior shelf rotation and that some type of contractile system was involved in the process. In order to discern the type of contractile system involved, antagonists were added and *in vitro* effects on anterior and posterior shelf rotation were measured.

When curare (2×10^{-3} M) was added to the medium, posterior movement was not impaired, implying that the skeletal muscle in region 1 was not involved as an agonist muscle in rotation. Interestingly, posterior shelf rotation was stimulated (Table 9.2). Also, a cross-sectional cut through the palate, two-thirds of the way along the anterior to posterior axis, before culture, causes a normal rotation of the anterior two-thirds of the palate; the posterior one-third, containing the skeletal muscle, rolls up in the opposite direction (L. Brinkley, personal communication). These observations suggest an antagonist role for region 1 skeletal muscle. However, more experiments need to be done to assess this possibility.

Next, atropine was tested for its effect on posterior shelf rotation; inhibition would imply an effect on muscarinic receptors of smooth muscle or non-muscle contractile systems (an analogy with sea urchin gastrulation[18]). Little effect could be measured (Table 9.2). The nicotinic antagonist, hexamethonium, was next tested. Inhibition of the posterior palate was profound, about 35% of control at 10^{-6} M and 10^{-4} M after 2 h in culture. No significant effect was observed at the anterior end of the palate. After overnight culture, 10^{-4} M hexamethonium almost completely suppressed posterior shelf rotation with no effect on the anterior end (Table 9.2). An effect on a nicotinic receptor implies that the inhibition was taking place at the terminus of a preganglionic fibre, possibly from the pterygopalatine ganglion which is located adjacent to region 2 cells. This situation is analogous to the adrenal medulla whereby preganglionic fibres regulate secretion of catecholamines from cells by release of acetylcholine. It should be pointed out that the adrenal medulla is an APUD type of cell[19]. Some APUD cells may possibly be regulated by tight-like junctions ('enpassant' synapses)[20]. Region 2 cells appear to have tight-like junctions with axons and contain dense core vesicles, clear and coated. Therefore, the intriguing possibility exists that region 2 cells are APUD cells, derived from the neural crest, which apparently migrate into the palate from outgrowth of the maxillary process.

Next, the effect of serotonin on palate shelf rotation was tested (Table 9.2). The predominant effect was a marked stimulation of rotation at the anterior end; with 10^{-5} M serotonin, movement in a 2 h culture was 319% of control values ($p < 5 \times 10^{-5}$)[21]. The serotonin antagonist, methysergide, at 10^{-4} M predominantly inhibited anterior shelf rotation: anterior, 12% of control ($p < 0.005$); posterior, 77% of control (not significant). Dose–response experiments were carried out with both serotonin and methysergide. The shape of the stimulation curve with serotonin and the inhibition curve with methysergide were biphasic[21]. Failure to obtain a simple dose–response pattern for these agents may be attributed to their paradoxical effects, as was observed in other systems: serotonin can act as an antagonist as well as an agonist[22]; methysergide shows a 'biphasic' effect on canine nasal circulation[23]; methysergide shows mixed agonist–antagonist activity of dopamine and serotonin receptors[24]. In addition, the serotonin antagonist, cyproheptadine (10^{-9} M), inhibited the anterior palate more than the posterior end *in vitro*. When injected twice (3.24 mg/kg) into pregnant dams at day 14.5 and 14.75, cyproheptadine inhibited palate rotation at day 15.5; anterior, $p < 5 \times 10^{-4}$; posterior, $p < 0.01$. Embryos were also incubated in culture for 2 h in the presence of 10^{-4} M methysergide and a morphological analysis of palate cells was compared to control cultures. Cells of region 3 present in anterior and midpalate failed to contract, whereas posteriorly, changes in region 2 cells were unaffected. These observations imply that the contraction of region 3 cells, which is assumed necessary for anterior shelf rotation, is under serotoninergic control.

Presence of serotonin in the palate

Since serotonin and its antagonists affect rotation in embryo culture, experiments were carried out to determine whether serotonin is associated with the palate before elevation[25]. First, dissected palates in culture incorporated ninefold more |³H|5HT (serotonin) at 35 °C than at 4 °C, implying that the palate contains an active uptake mechanism for serotonin. Secondly, the presence of the enzyme, 5HTP decarboxylase, was sought by incubating embryos or excised palates in cultures with |³H|5HTP. The presence of 5HTP decarboxylase in the palate was confirmed by measuring labelled |³H|5HT, separated by TLC. Palates were excised into three pieces (anterior, mid- and posterior) and incubated with |³H|5HTP. It was observed that mid-palate contained threefold more 5HTP decarboxylase than either the anterior or the posterior ends. This result supports the notion that some or all of the region 3 cells contain serotonin, since region 3 cells are found predominantly in mid-palate. Finally, serotonin content in the palate was measured during development. It was observed that the highest level of serotonin was found at day 14.5, the time just prior to elevation. This result also supports the role of serotonin in regulating palate morphogenesis.

Cell culture of mesenchyme

Palatal mesenchymal cells have been grown in dispersed monolayers or as explants. When the palate mesenchyme is explanted for two days in culture, small stellate cells migrate out of the explant which have a morphology strikingly similar to chick embryo neural crest cells growing out of neural tubes[26]. Similarly, many cells in this population have nerve-like morphology, with long cytoplasmic processes.

If the anterior and mid-palate were under serotoninergic control, it might be expected that some of these cells undergo contraction in the presence of serotonin or 5,6-dihydroxytryptamine (DHT), the non-metabolizable serotonin toxin. First, palatal mesenchyme was dispersed and plated as monolayers. 5HT or DHT (10^{-5} M) caused about 5–10% of the cells to contract.

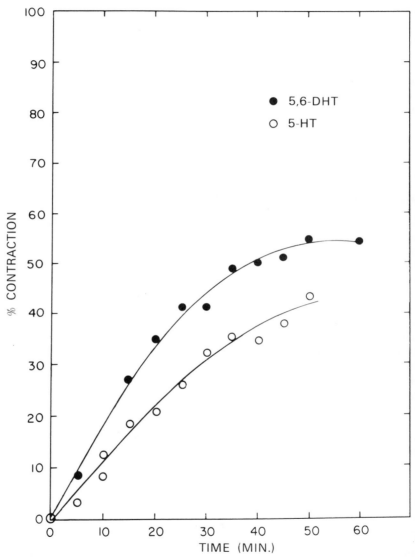

Figure 9.6 Effect of 5HT and 5,6-DHT on cell contractility. Two day explants were exposed to 10^{-5} M 5HT or 5,6-DHT. Every 5 min, stellate cells which migrated from explants were photographed. Determination of contraction was based on decrease in surface area as cells rounded up. Photographs were xeroxed, cells were cut out and weighed. Percentage contraction equals difference in weight between zero and observed time divided by zero time weight × 100

Next, palate mesenchyme was explanted for two days. When DHT or 5HT were added to the culture medium, and the stellate cells followed by phase microscopy, 70–90% of them underwent contraction. Cells contracted first in the short axis and then in the long axis. The contraction process appeared to be complete within 60 min (Figure 9.6). Some cells rounded-up to such an extent that they sloughed off the polylysine coated glass substratum. Since only specific cells responded to serotonin or DHT, this is evidence that the contraction process is not due to general cell toxicity. Finally, when the formaldehyde-induced fluorescence reaction was carried out, the stellate cells were unreactive whereas a cell population in 7-day explants showed a strong yellow fluorescence, characteristic for serotonin. It is possible that the serotonin reactive cells represent those of region 3 *in situ* and the stellate cells represent the rest of the peripheral mesenchymal cells.

In summary, these experiments suggest that neural crest-derived cells are found in the palate mesenchyme, which play a role in morphogenesis regulated by neurotransmitters[27].

DISCUSSION
Model of palate morphogenesis

Morphological studies using myosin ATPase histochemistry, electron microscopy, and immunofluorescence with actin and myosin antibodies have localized contractile systems within three regions of the palate (see Figures 9.1–4). Cells in the two non-muscle contractile systems (regions 2 and 3) contained a large number of filopodial processes with a high concentration of the contractile proteins, actin and myosin. These cells contain clear and dense core vesicles consistent with containing acetylcholine and serotonin.

Subsequent studies attempted to determine whether these cells played a role in palate morphogenesis. Not only did these studies support the notion that region 2 and 3 cells contracted to help elevate the palate, but that contraction of other peripheral cells aids this process. A model of the interactions of these cells is presented diagrammatically in the anterior (Figures 9.7a–c) and posterior (Figures 9.7d–f) palate. Anteriorly, contraction of region 3 cells (Figures 9.7a and b) produces a horizontalization of the oral epithelium. Subsequent alignment of medical and tongue side mesenchymal cells acts to flatten or draw in the tongue side of the shelf. The combination of these cellular contractions results in anterior palate rotation (Figure 9.7c). Posteriorly, prior to rotation, no orientation is seen in region 2 and no region 3 exists (Figure 9.7d). However, during rotation a contractile force is produced by cells of region 2 and the mesenchymal cells below this area. These contractions, assisted by the elongation and possible contraction of the tongue side epithelial cells, allow for the rotation of the posterior palate (Figure 9.7f). This model is based upon the cell shape changes summarized in Figure 9.5. The contraction of the appropriate peripheral mesenchymal cells, including those of regions 2 and 3, is supported by the observation that ATP caused a preferential contraction of these cells in glycerinated heads which was proportional to palate shelf rotation. Contraction of these cells by a cytoplasmic contrac-

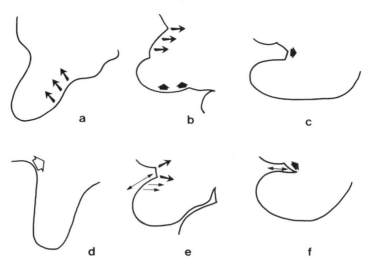

Figure 9.7 a–c Diagrammatic representation of anterior shelf elevation. a, At day 14.5, the contractile system in region 3 consists of elongate cells aligned perpendicular to the oral epithelium (large arrows). b, Rotation begins with contraction of region 3 cells (short arrows) elevating the oral epithelium to the horizontal position. This results in a bulging of the tongue side epithelium. Underlying this bulging epithelium and at the bend, the mesenchymal cells elongate and orient perpendicularly (long arrows). c, Contraction of the tongue side mesenchymal cells (short arrow) flattens or draws in the epithelium completing anterior shelf rotation

Figure 9.7 d–f Diagrammatic representation of posterior palate elevation. d, At day 14.5, no orientation is seen in region 2, although morphologically a contractile system exists (open arrow). e, Rotation begins with the orientation and elongation of region 2 cells (large arrows). The tongue side epithelial cells become elongated (two-headed arrow), as do the mesenchymal cells below region 2 (small arrows). f, Subsequent contractions of region 2 cells and the mesenchymal cells along the tongue side epithelium (short large arrow), plus possible epithelial contraction (short two-headed arrow) result in the elevation of the posterior palate

tile system is supported by the observation that cytochalasin B blocked the ATP-induced palate shelf rotation in glycerinated heads. Furthermore, involvement of the neurotransmitters, acetylcholine and serotonin, supports a contractile system in regulating palate morphogenesis. Further support comes from the observation that serotonin causes a contraction of stellate cells from mesenchymal explants of palate (Figure 9.6) and methysergide, the serotonin antagonist, blocked contraction of region 3 cells and palate rotation in embryo culture.

For the contractile systems in regions 2 and 3 and other peripheral mesenchymal cells to successfully produce a force capable of elevating the palate shelves, the 'origins' and 'insertions' of those systems should be defined. The mesenchymal contractile systems could be anchored by embedding cellular projections into the basement membrane of the tongue side and oral epithelium. These connections could serve as an 'insertion' for both systems. The cells of region 2 were found to abut the perichondrium of the cranial base, and during rotation were aligned from the epithelium to abut the cells of the ossification centre in region 2. These structures may serve as the 'origin' of the region 2 system. An 'origin' for region 3 is less easily defined. Observations of histological preparations of glycerinated palate have shown that all of the mesenchymal cells of the palate are connected in a web-like system by fine cellular projections, acting as a syncytium. These observations suggest that a contraction of the elongate cell system of region 3 could produce an intrinsic shelf force by 'originating' from the interior palatal mesenchyme.

Finally, the region 3 system on the oral side and the peripheral mesenchymal components on the medial and tongue side appear to be regulated by serotonin and contribute to anterior and mid-shelf rotation. On the other hand, acetylcholine appears to regulate posterior shelf rotation by its effect on region 2 cells on the tongue side mesenchyme. This seeming complexity can be resolved if one appreciates that the region 3 peripheral mesenchymal system is rather efficient at rotating the anterior and mid-portion of the shelf after the tongue has descended out of the way. On the other hand, the contractile system in the posterior end is present only on the tongue side and would seem inefficient, having to 'remodel' this part of the shelf. However, this places the motive force behind the tongue and helps extrude the palate shelf around the tongue which cannot descend at this posterior location. Dual regulation of these two contractile systems by serotonin and acetylcholine would seem to be an advantage in coordinating these complicated cell movements.

Teratological implications

It is becoming increasingly apparent that region 2 and 3 cells and other peripheral components are derived from neural crest cells which have migrated from the cranial neural tube. First, their location on the periphery of the mesenchyme adjacent to the epithelium is consistent with crest cells migrating along the craniofacial ectoderm[26]. Their ability to synthesize and/or respond to neurotransmitters by contraction is consistent with their origin from neural crest cells which are known to differentiate into nerves and neuroendocrine cells such as the chromaffin cells of he adrenal medulla[28,29].

It is known that CNS depressants such as alcohol, anticonvulsants, tranquillizers and barbiturates are teratogenic in experimental animals and can produce cleft palate[30-33]. Furthermore, it is thought that alcohol, anticonvulsants and tranquillizers also cause malformations in the human[34-37]. That these CNS depressants could cause a malformation such as cleft palate would seem paradoxical. However, if nerve-like cells (derived from the neural crest) play a role in palate morphogenesis, then a specific interaction with these cells by CNS depressants would seem feasible. Binding to specific membrane receptors could be a feature of this interaction. The net result could be a delay or blockade in palate rotation due to the toxic effect on the mesenchymal cells derived from the neural crest. It would seem possible to test this hypothesis in the future.

ACKNOWLEDGEMENTS

This work was supported by a research grant from the National Institute of Dental Research (DEO3469) and a Center Grant in Mental Retardation (HDO5221).

References

1. Walker, B. E. and Fraser, F. C. (1957). The embryology of cortisone-induced cleft palate. *J. Embryol. Exp. Morphol.*, 5, 201
2. Faludi, G., Gotlieb, J. and Meyers, J. (1966). Factors influencing the development of steroid myopathies. *Ann. N.Y. Acad. Sci.*, 138, 61
3. Walker, B. E. and Fraser, F. C. (1956). Closure of the secondary palate in three strains of mice. *J. Embryol. Exp. Morphol.*, 4, 176
4. Lessard, J. L., Wee, E. L. and Zimmerman, E. F. (1974). Presence of contractile proteins in mouse fetal palate prior to shelf elevation. *Teratology*, 9, 113
5. Pollard, T. D. and Weihing, R. R. (1974). Actin and myosin in cell movement. *Crit. Rev. Biochem.*, 2, 1
6. Babiarz, B. S., Allenspach, A. L. and Zimmerman, E. F. (1975). Ultrastructural evidence of contractile systems in mouse palates prior to rotation. *Dev. Biol.*, 47, 32
7. Padykula, H. A. and Herman, E. (1955). Factors affecting the activity of adenosine triphosphatase and other phosphatases as measured by histochemical techniques. *J. Histochem. Cytochem.*, 3, 161
8. Babiarz, B. S., Kuhn, E. M. and Zimmerman, E. F. (1976). Embryonic contractile system in mouse palate during development. *Teratology*, 13, 15A
9. Kuhn, E. M., Babiarz, B. S., Lessard, J. L. and Zimmerman, E. F. (1977). Palate morphogenesis. I. Localization of non-muscle contractile systems by immunofluorescence and electron microscopy. Presented at the *5th Int. Conf. on Birth Defects*, Acta Medica Fndn., Aug. 21-27, Montreal
10. Krawczyk, W. S. and Gillon, D. G. (1976). Immunofluorescent detection of actin in non-muscle cells of the developing mouse palatal shelf. *Arch. Oral Biol.*, 21, 503
11. Innes, P. B. (1978). The ultrastructure of the mesenchymal element of the palatal shelves of the fetal mouse. *J. Embryol. Exp. Morphol.*, 43, 185
12. Wee, E. L., Wolfson, L. G. and Zimmerman, E. F. (1976). Palate shelf movement in mouse embryo culture: Evidence for skeletal and smooth muscle contractility. *Dev. Biol.*, 48, 91
13. Szent-Györgyi, A. (1951). *The Chemistry of Muscle Contraction*, 2nd Ed. (New York: Academic Press)
14. Norberg, B. (1970). Amoeboid movements and cytoplasmic fragmentation of glycerinated leukocytes induced by ATP. *Exp. Cell Res.*, 59, 11
15. Wee, E. L., Brinkley, L. and Zimmerman, E. F. (1977). ATP-induced palate shelf rotation in glycerinated mouse heads and inhibition by cytochalasin B. *Teratology*, 15, 22A
16. Pollard, T. D. (1976). The role of actin in the temperature-dependent gelation and contraction of extracts of *Acanthamoeba. J. Cell Biol.*, 68, 579
17. Stossel, T. P. and Hartwig, J. H. (1976). Interactions of actin, myosin and a new actin-binding protein of rabbit pulmonary macrophages. II. Role in cytoplasmic movement and phagocytosis. *J. Cell Biol.*, 68, 602
18. Gustafson, T. and Toneby, M. (1970). On the role of serotonin and acetylcholine in sea urchin morphogenesis. *Exp. Cell Res.*, 62, 102
19. Pearse, A. G. E. and Polak, J. M. (1974). Endocrine tumors of neural crest orgin: Neurolophomas, apudomas and the APUD concept. *Med. Biol.*, 52, 3
20. Welsch, U. and Schubert, C. (1975). Observations on the fine structure, enzyme histochemistry, and innervation of parathyroid gland and ultimobronchial body of *Chthonerpeton indistinctum.* (Gymnophiona, Amphibia). *Cell Tissue Res.*, 164, 105
21. Wee, E. L., Babiarz, B. S., Zimmerman, S. and Zimmerman, E. F. (1977). Palate morphogenesis. III. Effects of serotonin and its antagonists on rotation in embryo culture. Presented at the *5th Int. Conf. on Birth Defects.* Acta Medica Fndn., Aug. 21-27, Montreal
22. Allen, G. S., Gross, C. J., Henderson, L. M. and Chou, S. N. (1976). Cerebral arterial

spasm. Part 4: *In vitro* effects of temperature, serotonin analogues, large non-physiological concentrations of serotonin and extracellular calcium and magnesium on serotonin-induced contractions of the canine basilar artery. *J. Neurosurg.*, **44**, 585
23. Schönbaum, E., Vargaftig, B. B., Lefort, J., Lamar, J. C. and Hasenack, T. (1975). An unexpected effect of serotonin antagonists on the canine nasal circulation. *Headache*, **15**, 180
24. Ferrari, C., Paracchi, A., Rondena, M., Beck-Peccoz, P. and Faglia, G. (1976). Effect of two serotonin antagonists on prolactin and thyrotrophin secretion in man. *Clin. Endocrinol.*, **5**, 575
25. Zimmerman, E. F. and Roberts, N. (1977). Palate morphogenesis. IV. Synthesis of serotonin. Presented at the *5th Int. Conf. on Birth Defects*, Acta Medica Fndn., Aug. 21–27, Montreal
26. Horstadius, S. (1950). *The Neural Crest.* p. 111 (London and New York: Oxford Univ. Press)
27. Kujawa, M. J. and Zimmerman, E. F. (1978). Palate cells in culture: 5HT-induced contraction and neural crest origin. *Teratology*, **17**, 29A
28. Biales, B., Dichter, M. and Tischler, A. (1976). Electrical excitability of cultured adrenal chromaffin cells. *J. Physiol.*, **262**, 743
29. Brant, B. L., Hagiwara, S., Kidokoro, Y. and Miyazaki, S. (1976). Action potentials in the rat chromaffin cell and effects of acetylcholine. *J. Physiol.*, **263**, 417
30. Kronick, J. B. (1976). Teratogenic effects of ethyl alcohol administered to pregnant mice. *Am. J. Obstet. Gynecol.*, **124**, 676
31. Gibson, J. E. and Becker, B. A. (1968). Teratogenic effects of diphenylhydantoin in Swiss-Webster and A/J mice. *Proc. Soc. Exp. Biol. Med.*, **128**, 905
32. Miller, R. P. and Becker, B. A. (1975). Teratogenicity of oral diazepam and diphenylhydantoin in mice. *Toxicol. Appl. Pharmacol.*, **32**, 53
33. Walker, B. E. and Patterson, A. (1974). Induction of cleft palate in mice by tranquilizers and barbiturates. *Teratology*, **10**, 159
34. Jones, K. L., Smith, D. W., Ulleland, C. N. and Streissguth, A. P. (1973). Pattern of malformation in offspring of chronic alcoholic mothers. *Lancet*, i, 1267
35. Hanson, J. W. and Smith, D. W. (1975). The fetal hydantoin syndrome. *J. Pediatr.*, **87**, 285
36. Safra, M. J. and Oakley, G. P. (1975). Association between cleft lip with or without cleft palate and prenatal exposure to diazepam. *Lancet*, ii, 478
37. Saxén, I., and Saxén, L. (1975). Association between maternal intake of diazepam and oral clefts. *Lancet*, ii, 498

10
Alterations in macromolecular synthesis related to abnormal palatal development

R. M. PRATT, A. A. FIGUEROA, R. M. GREENE,
ANN L. WILK AND D. S. SALOMON

SECONDARY PALATE DEVELOPMENT
Introduction

Closure of the secondary palate in most vertebrates involves reorientation of the bilateral palatal shelves from a vertical position on either side of the tongue to a horizontal position above the tongue (Figure 10.1). The apposing medial-edge epithelia (MEE) adhere and then degenerate, allowing the merging of mesenchymal cells to form a single continuous structure, the secondary palate. The overall process of palatogenesis can be divided into four stages: growth of the vertical palatal shelves, reorientation to a horizontal position and, finally, epithelial adhesion and subsequent programmed epithelial cell death. These events occur at midgestation after most organogenetic events have occurred in the embryo.

A large number of teratogenic agents have been shown to be capable of inducing isolated cleft palate when given at various periods during gestation[1]. This list ranges from such agents as X-rays to chemicals, such as β-aminoproprionitrile, whose mechanism of action is known and limited to tissues where it interferes with collagen cross-linking. The diverse nature of these and other agents suggests that a simple mechanism for producing cleft palate does not exist. It is more likely that these agents operate by perturbing different morphogenetic events at specific stages of palatal development.

Development of the secondary palate is a complex morphogenetic process which is dependent on certain biochemical events occurring both within the shelf and in other surrounding structures such as the tongue and mandible. This chapter will describe the important macromolecules synthesized by the palate during development and the possible role that these components may perform during palatal development. The importance of these molecules will

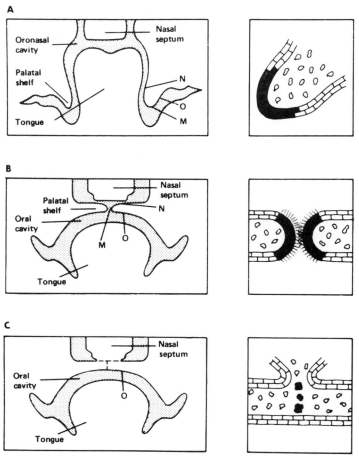

Figure 10.1 Schematic diagrams of secondary palate development. Schematic frontal sections through the anterior rodent head at the following days of gestation: A, Day 15 rat, Day 13 mouse; B, Day 16 rat, Day 14 mouse; C, Day 17 rat, Day 15 mouse. The schematics of the palatal shelves to the right show that the cells of the medial-edge (M/shaded area) cease DNA synthesis (A) undergo glycoconjugate-mediated ($\approx$) adhesion (B) and finally cell death (C).

This sequence of events does not occur in the cells of the oral (O) or nasal (N) epithelium

be illustrated by alteration in their synthesis or degradation following the administration of various teratogenic agents. Particular emphasis will be placed on the biochemical basis of gene-environment interactions such as the production of cleft palate by cortisone in sensitive strains of mice.

Glucocorticoid effects on proliferation of secondary palate cells

Glucocortoids participate in the development or growth of embryonic cells and tissues[2,3]. The appearance of specific cytoplasmic glucocorticoid receptors in fetal tissues at defined periods during development has been implicated as one factor which might determine the onset of hormonal responsiveness in these tissues to glucocorticoids[4-7]. Alternatively, receptors for these hormones may already be present in embryonic or fetal tissues about to undergo differentiation, but these tissues may be incapable of responding to hormones due either to an absence of specific nuclear receptor or acceptor proteins for the glucocorticoids (Figure 10.2) or to an unavailability of appropriate hormone levels in the mother and/or fetus to trigger such events.

Various inbred strains of mice exhibit different degrees of susceptibility to glucocorticoid-induced cleft palate[8,9]. A/J mice treated with glucocorticoids between days 11 and 14 of gestation produce 100% offspring with cleft palate, whereas similarly-treated C57BL/6J (C57) mice produce only 20–25% offspring with cleft palate. Both maternal and embryonic factors probably contribute to the strain-dependent susceptibility to glucocorticoid-induced cleft palate[10-13], and the reasons for the increased susceptibility of certain strains of mice to cleft palate production are becoming clear.

Glucocorticoids can traverse the rodent placenta without being metabolized[14,15]. Furthermore, although glucocorticoids are synthesized and metabolized in the maternal adrenals and liver in a manner which quantitatively

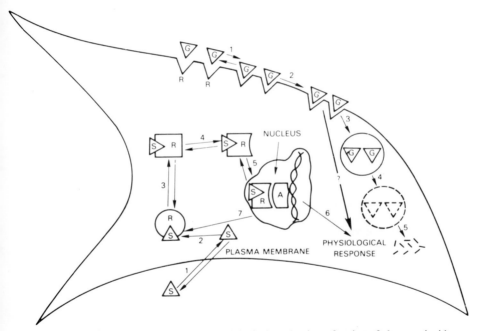

Figure 10.2 Schematic diagram of the physiological mechanism of action of glucocorticoids and epidermal growth factor (EGF). Steroid hormones (S) diffuse freely across the plasma membrane (1), bind specifically to cytoplasmic receptor proteins (2) which then undergo conformational change (3, 4). The activated hormone-receptor complex (SR) enters the nucleus (5), binds to acceptor proteins (A) associated with the chromatin resulting in increased synthesis of specific mRNA molecules (6). In contrast, epidermal growth factor (G) binds to cell surface glycoproteins (1), followed by patching (2), internalization (3) and lysosomal-mediated degradation (4, 5). EGF may exert its effect on proliferation by some action at the cell surface

differs among various strains of mice[16-19], unmetabolized glucocorticoids in the mouse fetus are the active teratogenic agents. The differential distribution and binding of radioactive glucocorticoids to fetal tissues has been shown to be correlated with the *in vivo* strain-dependent susceptibility to steroid-induced cleft palate[20-23].

Variations in maternal plasma or fetal corticosterone levels during midgestation could also contribute to normal and abnormal development of the secondary palate in the mouse. The concentrations of maternal plasma corticosterone are six to ten-fold higher in A/J, C57, and ICI mice during days 11 −15 of gestation than those of non-pregnant animals[24, 25]. Although A/J mice are more sensitive to the teratogenic effects of exogenously administered glucocorticoids than C57 mice in regard to cleft palate production, no significant quantitative differences in the endogenous concentration of maternal plasma corticosterone can be detected between these two strains[24]. However, administration of pharmacological doses of glucocorticoids to pregnant mice during midgestation is known to substantially elevate the already high levels of plasma corticosterone in the maternal circulation[26]. Therefore, there may well be differences in the maternal metabolism of high doses of steroids which could result in a difference in the final plasma concentration of steroids between these two strains. In contrast to the high levels of corticosterone in the maternal plasma, the levels of corticosterone found in both A/J and C57 fetuses during this same period are lower and do not exhibit any marked fluctuation between days 11 and 13 in parallel with changes in the maternal plasma levels. The absence of significant differences in the concentration of maternal or fetal corticosterone between A/J and C57 mice suggests that the *in vivo* strain differences in glucocorticoid sensitivity are not related to any intrinsic differences between these two strains in the endogenous levels of corticosterone during days 11 to 13 of gestation. Nevertheless, the elevation in maternal corticosterone in both strains between days 11 and 13, and the subsequent elevation in fetal corticosterone after day 13, may be related to certain stages of normal palatogenesis. In this respect, fusion of the developing palatal shelves occurs late on day 14 or early on day 15 of gestation. Whether this rise in both maternal and fetal corticosterone is entirely coincidental or whether changes in glucocorticoid levels during this period might serve to initiate developmental events in the secondary palate remains to be determined[27].

More recent studies have shown that mouse embryonic facial and palatal mesenchyme cells obtained from day 14 embryos possess high affinity glucocorticoid receptor proteins which bind dexamethasone. Salomon and Pratt[28] have shown that A/J mesenchyme cells, either freshly isolated or maintained *in vitro* as primary cultures, possess approximately two to three times more receptor proteins than C57 mesenchyme cells. Extension of similar studies to other strains of mice has also shown that there is a reasonable correlation between the *in vitro* levels of cytoplasmic glucocorticoid receptor proteins in maxillary mesenchyme cytosols and the *in vivo* degree of sensitivity of these strains to glucocorticoid-induced cleft palate (see Table 10.1 and Salomon and Pratt[29]). Although there is a quantitative difference in the level of cytoplasmic receptor proteins between A/J and C57 maxillary or palatal mesenchyme

Table 10.1 Comparison of maxillary cytoplasmic glucocorticoid receptors and cleft palate frequency

Strain	fmoles of [^{3}H]dexamethasone specifically bound/mg cytosol protein*	% Cleft palate†
SWR/FR	273	100
SWR/NIH	200	100
DBA/IJ	95	92
A/J	53	100
CBA/J	35	12
C57BL/6J	23	19

* Values from three to six determinations obtained from Scatchard plots of the specific binding of [^{3}H]dexamethasone (1–100 nM) to day 14 palatal processes. (See Salomon and Pratt[28-29])
† Animals were administered (s.c.) 2.5 mg cortisone acetate on days 11–14 of gestation

cells, no quantitative differences were observed between these two strains in either the ability of these cytoplasmic receptors to translocate to nuclear acceptor sites or the actual number of nuclear receptor sites. Similar strain specific differences in cytoplasmic glucocorticoid receptor levels have been demonstrated after measuring hydrocortisone binding in mouse fetal palates[30].

The presence of glucocorticoid receptors in both the palate and surrounding facial mesenchyme cells suggests that these cells have the potential to respond biologically to glucocorticoids. Moreover, quantitative differences in the number of glucocorticoid receptors between A/J and C57 cells suggest that such differences might be reflected in the magnitude of the biological responses induced by glucocorticoids in these two cell types.

In a recent study, the effect of glucocorticoids on primary cultures of mouse embryonic facial mesenchyme cells has been studied *in vitro*[29]. Glucocorticoids promote a reduction in cell number and a simultaneous decrease in the incorporation of [^{3}H]thymidine into DNA in both A/J and C57 mesenchyme cells. No significant changes in total protein synthesis were observed in either cell type following the administration of glucocorticoids. In both A/J and C57 cultures, this reduction in cell number exhibits a dose-dependent response to dexamethasone, is specific for glucocorticoids, and is dependent upon the concentration of serum in which the cells are maintained. A/J cells show a greater sensitivity to the inhibitory effect of the steroids on cell growth than comparably-treated C57 cells. This difference appears to be due to the higher level of glucocorticoid receptors in these cells. In both cell types, a correlation exists between the degree of growth inhibition or reduction of thymidine incorporation and the level of glucocorticoid receptors. Genetic differences in relation to steroid responses previously identified in established cell lines *in vitro*[31,32] may have similar counterparts *in vivo* in various inbred strains of mice which exhibit differential susceptibility to glucocorticoid-induced cleft palate. This differential steroid effect on [^{3}H]thymidine incorporation in primary cultures can also be observed in organ cultures of palatal shelves from A/J and C57 mice which have been maintained in the presence of dexamethasone for 24 h (Table 10.2). There is a greater inhibition of thymidine incorporation in A/J secondary palates than in C57 palates, which correlates

Table 10.2 Correlation between binding of [³H]dexamethasone and inhibition of thymidine incorporation in day 13 A/J and C57BL/6J palates

Strain	fmoles of [³H]dexamethasone specifically bound/mg protein*	Inhibition of [³H]thymidine incorporation†
A/J	200	38.4%
C57BL/6J	76	10.4%

* Day 13 palates (freshly dissected) were incubated at 37 °C in Dulbecco's modified Eagle's medium containing 0.1% BSA with 8×10^{-8} M [³H]dexamethasone in the absence or presence of nonradioactive dexamethasone (5×10^{-5} M) to correct for nonspecific binding for 45 min. Palates were then washed extensively at 4 °C with PBS, solubilized and counted

† Day 13 palates were maintained in organ culture in growth medium $\pm\ 10^{-6}$ M dexamethasone. After 6 h of incubation at 37 °C, [³H]thymidine (5 μCi/ml) was added to the cultures. At 24 h of culture, the palatal shelves were processed for incorporation of [³H]thymidine into TCA-insoluble macromolecules. Incorporation was normalized to a per mg protein basis and data is expressed as the percent inhibition of incorporation in steroid-treated cultures

with the higher level of glucocorticoid receptors measured in whole palatal shelves of A/J mice. Autoradiographic analysis (Pratt and Salomon, unpublished observations) demonstrated that the inhibition of thymidine incorporation by dexamethasone is more apparent in the palatal mesenchyme cells than in the nasal, oral or medial-edge epithelial cells. These results are in accord with the observations that dexamethasone and cortisone preferentially inhibit *in vivo* the proliferation of mesenchyme cells of the palatal processes in rats and mice[33-36].

Since low physiological doses of glucocorticoids stimulate maxillary mesenchyme cell growth *in vitro* and promote thymidine incorporation[29], and since the growth inhibitory effect of high concentrations of glucocorticoids is ameliorated by elevated concentrations in serum, it is conceivable that glucocorticoids either alone or through interactions with other hormones or growth factors may function to control certain stages of normal palatogenesis. It is noteworthy that dexamethasone may enhance cell growth, in conjunction with FGF, EGF or insulin, by modulating either the affinity or number of receptors for these mitogens[37,38]. Conversely, the production of cleft palate by pharmacological doses of glucocorticoids may be due to abnormally high levels of steroid in the fetus. Failure of the bilateral shelves to make adequate contact and eventually fuse at the appropriate developmental stage may be due to a reduction in size of the shelves as a result of inhibition of mesenchymal cell proliferation. Likewise, glucocorticoids might reduce the production of extracellular matrix components in the developing palatal shelves such as hyaluronic acid, glycosaminoglycans or collagen[39-41] or control the composition of cell surface components[42,43].

Cellular proliferation and extracellular matrix

A motive shelf 'force' developing within the shelves has been proposed by Walker and Fraser[44] as one factor which might account for palatal shelf elevation. The process of elevation occurs rapidly in the rodent and it is unlikely that increased proliferation of palatal mesenchyme cells, which constitute 80% of the total palatal cells, could be responsible for shelf rotation. In fact, various studies have shown that the cellular proliferation in the palatal mesenchyme decreases during elevation[10] and maximal cell proliferation appears to precede elevation by 24 to 48 h and is highest in the mesenchyme underlying the medial-edge epithelium[33]. Even though proliferation does not appear to play a direct role in elevation, it is evident that a reduction in mesenchymal cell proliferation at any time preceding or following elevation could serve to reduce the size of the shelves, therby delaying or preventing contact of the opposing shelves.

The palatal shelves synthesize and accumulate sulphated proteoglycans and hyaluronate prior to and during elevation[45]. Hyaluronate is an extremely hydrated, high molecular weight glycosaminoglycan which is normally associated with morphogenetic movements and tissue swelling[46]. The accumulation of hyaluronate may be especially pronounced in the mesenchyme underlying the presumptive oral epithelium of the palate which appears to contain more extracellular space and glycosaminoglycans than the mesenchyme beneath the nasal epithelium where the mesenchyme cells are more densely packed. A recent study by Wilk et al.[47] has shown that there is a good correlation between cleft palate production in rats by chlorcyclizine and its ability to promote degradation of hyaluronate. Hyaluronate may therefore play a key role in palatal elevation by imparting a certain degree of fluidity to the mesenchyme which is presumably necessary for tissue reorientation occurring at elevation.

Collagen rapidly accumulates in the palate during development and accounts for approximately 5% of the total palatal protein[48]. Collagen fibrils can be observed throughout the palatal mesenchyme, but are particularly abundant in the area adjacent to the basal lamina of the oral epithelium[49]. Preliminary immunofluorescent localization experiments indicate that cartilage-specific Type II collagen is absent in the palate, Type IV is restricted to the basement lamina, Type III and Type I are found in the palatal mesenchyme. BAPN appears to delay shelf elevation and cause cleft palate by preventing the cross-linking of embryonic collagen[50]. This suggests that the structural stability of collagen may be important for the biochemical events during elevation. A more detailed description of the interaction of various extracellular matrix macromolecules can be found in a review by Greene and Pratt[51].

Epithelial cell adhesion and programmed cell death

Initial adhesion of the horizontal homologous palatal shelves may be explained in part by the presence of an extracellular glycoconjugate-rich surface coat on the medial-edge epithelium. Apposing epithelial cell membranes of the bilateral palatal shelves are separated by a 10–20 nm glycoconjugate-filled space[52] (Figure 10.1). Initial adhesion presumably serves to keep the shelves in contact until more permanent desmosomal connections can be established between the medial-edge epithelial cells (Morgan and Pratt, unpublished). A glycoconjugate (glycosaminoglycans, glycoproteins and glycolipids)-rich surface coat dramatically appears on the epithelial surface (MEE) prior to contact[53,54]. Diazo-oxo-norleucine (DON), a glutamine analogue, has been shown to prevent adhesion between homotypic palatal shelves *in vitro*[55]. Shelves cultured in the presence of DON exhibit a substantial reduction in the binding of concanavalin A at the medial-edge epithelial surface. The biochemical nature of the surface coat is complex but recent studies (Pratt, unpublished) indicate the presence of two major cell surface glycoproteins (MW 115K and 190K) on the MEE. It is noteworthy that the cell surface glycoprotein, known as fibronectin (CSP, LETS) (220 K) which is the major cellular adhesive glycoprotein of fibroblasts in culture[56], is found on the palatal mesenchyme cells (Figueroa, Silver and Pratt, unpublished) but not on the nasal, oral or medial-edge epithelium.

The palatal medial-edge epithelium (MEE) in the mammalian secondary palate undergoes a sequence of biochemical events leading to cellular degeneration[57]. These events include cessation of DNA synthesis[58,59], increased synthesis of cell surface glycoconjugates, synthesis of lysosomal enzymes[60,61] and eventual MEE autolysis. This degeneration of the MEE allows the mesenchymal component of the palatal shelves to become continuous and form the definitive secondary palate (Figure 10.1). The MEE undergoes autolysis in organ cultures of isolated palatal shelves, indicating that contact with the apposing homologous palatal shelf is not necessary for degeneration[62-65]. Studies utilizing various inhibitors of macromolecular synthesis *in vitro* have demonstrated that MEE cell death is not a passive event but an active process requiring the programmed appearance of specific proteins[66]. Degeneration of the MEE *in vitro* can be prevented by DON, which appears to inhibit cell death by interfering with the glycosylation of glycoproteins. This alteration in glycosylation does not interfere with synthesis of lysosomal enzymes, but results in altered cellular distribution of lysosomal enzymes[67]. Acid phosphatase activity is normally present throughout the cytoplasm and autophagic vacuoles in the degenerating MEE cells, whereas in non-degenerating DON-treated MEE cells the enzyme activity is restricted to the Golgi and primary lysosomes. This and other data suggest that glycosylation may be important for the fusion of lysosomes with other intracellular components and/or for the eventual release of enzymes from the lysosomes at the final stages of cellular degeneration.

Increased, but transient, levels of palatal cyclic AMP have been found in the palatal shelves prior to and during fusion (Figure 10.3) suggesting that this intracellular secondary messenger may be involved in cell death of the MEE cells[59,68]. The regulatory role of cyclic AMP in cell function in various cell types is well documented[69]. Intracellular cyclic AMP levels are controlled in part by the activity of membrane associated adenylate cyclase. The appearance of adenylate cyclase in the MEE is time dependent, with demonstrable cytochemical activity only being detected just prior to and during MEE degeneration[68]. Enzyme activity is localized at the plasma membrane of degenerating MEE cells, suggesting that elevated levels of cyclic AMP and adenylate cyclase activity may be either directly involved in the initiation of cell death or a manifestation of cellular autolysis. Little information is available concerning the nature of the presumed hormonal agent(s) which might activate adenylate cyclase, although it has been suggested that β-adrenergic agents may be involved[70].

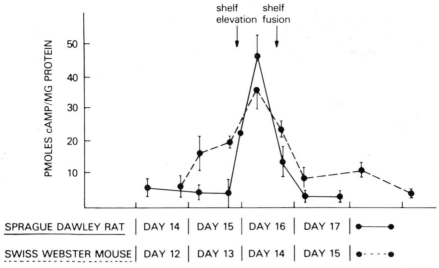

Figure 10.3 Cyclic AMP levels during secondary palatal development. The levels of cyclic AMP in whole palatal shelves (mesenchyme plus epithelium) were determined in the Swiss Webster mouse and Sprague-Dawley rat at various times during gestation (see Greene and Pratt[68])

Growth factors undoubtedly play a role in mammalian development and a great deal of work has been done on a mesenchyme factor which is capable of stimulating DNA synthesis in epithelial cells of the developing pancreas[3]. This factor appears to work in part through the cyclic AMP system. Epidermal growth factor (EGF), a small peptide (MW 6000) which is found in large amounts in the male mouse submandibular gland[71], has some interesting effects on MEE *in vitro*. The addition of EGF to palatal shelves in organ culture can prevent the normal degeneration of the palatal MEE[72]. This effect of EGF can be prevented by the simultaneous addition of dibutyryl cyclic AMP as shown by Hassell and Pratt[73]. Since this effect is specific for EGF, it suggests that cessation of DNA synthesis and cell degeneration may be related to the lack of intrinsic EGF in these MEE cells. EGF may, therefore, be required for proliferation of embryonic epithelial cells such as found in oral, medial edge and nasal epithelia of the palate.

PRIMARY PALATAL DEVELOPMENT

The primary palate is an embryonic facial structure that gives rise to the upper lip and maxillary bone which contains the four upper incisors. Embryonic development of the primary palate is a complicated morphogenetic event susceptible to alterations at a number of critical steps in its formation, leading to uni- or bilateral cleft lip with or without cleft palate.

Most of the studies on normal and abnormal primary palate development have been conducted with experimental animal model systems (such as rodents) which display similar facial development to the human. Formation of the primary palate in the rat occurs on days 11–13 of gestation (plug day 0). This process begins with the formation of the nasal placodes, which are ectodermal thickenings on both sides of the developing face (Figure 10.4).

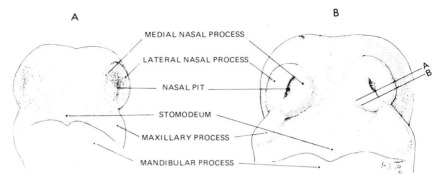

Figure 10.4 Schematic diagram of primary palate development. Schematic illustrations of rat embryos at two different stages of primary palate development. In Part A, a frontal view of the rat head is shown at $11\frac{1}{2}$ days of gestation, depicting the shallow invagination of the nasal pit and early development of the nasal processes. Part B demonstrates a later stage of primary palate development (day $12\frac{1}{2}$). The nasal processes are fully developed and fusion of the processes is progressing. Lines A and B indicate the section planes for Figure 10.5

Invagination of the nasal placode epithelium[74, 75] initiates the formation of the nasal pit. Subsequent proliferation of the surrounding mesenchyme (of neural crest origin) gives rise to the maxillary, median and lateral-nasal processes[76], and further growth of these processes encloses the nasal pit (Figure 10.4). The medial–nasal and maxillary processes merge first at the most caudal portion of the nasal pit, followed by epithelial contact and fusion of the lateral and median–nasal processes. Contact of the epithelial surfaces of these processes creates a temporary epithelial structure, the nasal fin, which subsequently undergoes autolysis and phagocytosis. Degeneration of the most posterior portion of the nasal fin serves to connect the nasal pit with the primitive oral cavity, thus establishing the primary palate.

Morphological observations suggest that the growth and fusion of facial processes are critical events which occur during primary palate formation. Evidence from studies with avian and rodent embryos demonstrated that the early phase of primary palate formation involves migration of cells derived from the cranial neural crest[76]. Rapid and differential proliferation of these mesenchyme cells, relative to the adjacent head mesenchyme cells[77] and concomitant matrix synthesis, results in growth of the facial processes. Distinct variations exist in the direction in which the facial processes grow in different strains of mice (A/J, C57/B6J and CL(FR)) which have different degrees of spontaneous cleft lip[78, 79]. These studies suggest that facial shape in these strains may be an important genetic predisposition for cleft lip. Whether or not an inherent difference exists in the rate of growth of the primary palate process in these strains remains to be determined. Administration of certain drugs during rodent development can produce delayed or retarded growth of the facial processes, resulting in facial clefts. For example, hadacidin affects growth of the maxillary process in the rat[80], 6-aminonicotinamide decreases growth of the medial-nasal process in A/J and CL/FR mice[81], and diphenylhydantoin reduces the size of the lateral-nasal process in the A/J mice[82]. These drugs probably operate through different mechanisms to reduce growth of the facial processes and give rise to facial clefts. It will be of interest to determine

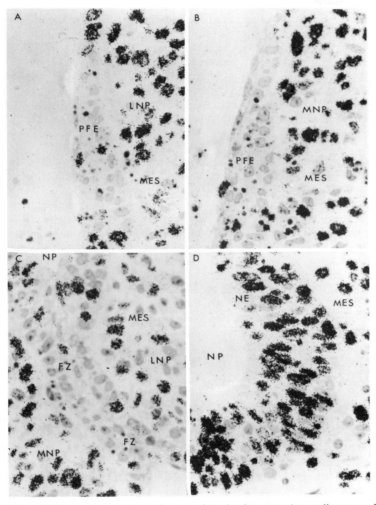

Figure 10.5 DNA synthesis during primary palate development. Autoradiograms of nasal processes from day $12\frac{1}{2}$ rat embryos exposed to [³H]thymidine *in vitro* (see Figueroa and Pratt[83]). In (A) and (B), significantly less labelling is observed in the PFE of the lateral and medial–nasal processes, compared to the underlying mesenchyme or adjacent nasal epithelium (D). Few cells of the fusion zone (C) are labelled when compared to the underlying mesenchyme. In A, B and D the plane of section is at Line A from Figure 10.4 and for C the plane of section is at Line B. LNP = lateral nasal process; MNP = medial–nasal process; PFE = presumptive fusion zone; NE = nasal epithelium; MES = mesenchyme; NP = nasal pit; FZ = fusion zone. × 774

the molecular basis for the differential sensitivity of these facial processes to these drugs.

During days 12 and 13 of gestation in the rat, fusion of the opposing facial processes occurs, involving specific cellular and biochemical events. (Figure 10.4). Recently it has been demonstrated autoradiographically that epithelial cells of the presumptive fusion zone of the facial processes selectively follow a series of events prior to contact and fusion[83]. These events include cessation of DNA synthesis 6 h prior to contact (Figure 10.5) with a concomitant increase in glycoprotein synthesis by these post-mitotic cells. The active synthesis of proteins and glycoproteins in these cells correlates well with the acquisition of ruthenium red staining[84] and Con A binding[74] shown to be present on the surface of the presumptive fusion epithelium of mouse embryos. These observations suggest that the cells of the presumptive fusion epithelium are 'programmed' to cease DNA synthesis prior to contact and actively secrete cell surface glycoconjugates, which may be involved in the initial steps of adhesion.

As to the possible mechanisms involved in the control of these 'programmed' events, recent studies[85] indicate the possible regulatory role of cyclic nucleotides. Previous studies have shown that cyclic AMP is a determinant molecule in cell differentiation, associated with reduced cell proliferation[69]. In the rat primary palate, levels of cyclic AMP transiently increase during the time in which the process of fusion is proceeding most actively[85].

After fusion of the opposing epithelial surfaces of the facial process, autolytic breakdown of the epithelial band (nasal fin) is necessary to achieve mergence and definitive mesenchymal union of the facial process. There is evidence that lysosomal enzymes are involved in this process as well as phagocytosis of the degenerating epithelial cells by macrophages[84].

It is of interest to note the clear similarity in developmental mechanisms involved during primary and secondary palate formation, structures which develop at quite different stages during embryonic development.

SUMMARY

The major purpose of this chapter was to review in some detail the recent evidence for alteration in macromolecular synthesis or function as a prime causative factor in teratogen induced cleft lip and palate. We have reviewed recent work which demonstrates a great deal of similarity between the major biochemical events occurring during primary and secondary palatal development. These similarities include the neural crest origin of mesenchyme, growth of the processes and mechanism of fusion. The major difference lies in the lack of a major morphogenetic movement during primary palate development, similar to that which occurs during secondary palatal shelf elevation and remodelling. On the other hand, the correct positioning of the three primary palatal processes for fusion is critical and particularly sensitive to interference.

Our knowledge of the biochemical events and regulatory processes is increasing rapidly, especially for the secondary palate. A major advance we have made in this area affords a partial explanation for the long-sought-after basis of the genetic differences in glucocorticoid-induced cleft palate between various strains of mice. We now know that the presence of higher levels of cytoplasmic steroid receptors serves in part to predispose the sensitive strains to a greater inhibition of craniofacial growth by cortisone and, therefore, increased frequency of cleft palate. It is also becoming clear that various hormones (glucocorticoids), growth factors (EGF) and cyclic nucleotides are critical for normal development, and that excesses or deficiencies are likely to lead to malformations.

Although the precise mechanism of elevation still remains to be determined, sufficient proliferation of palatal mesenchyme is essential for both proper shelf growth and the accumulation of extracellular macromolecules such as collagen, proteoglycans, glycosaminoglycans, contractile proteins and glycoproteins, all of which are thought to be necessary for some aspects of elevation and fusion. This does not rule out a contribution by adjacent craniofacial structures, such as the tongue and mandible, which are thought to play an important role in providing a proper structural environment for the developing shelves.

The changes occurring in the medial-edge epithelium throughout palatal development appear to represent a highly coordinated and programmed series of events eventually culminating in death of these cells. There are many agents which interfere with this cellular differentiation *in vitro* and our knowledge of the controlling factors, such as hormonal activation of the cyclic nucleotide second messenger system, is increasing. Further study and definition of the activity of these macromolecules during palate development and the manner in which they are regulated will help us to understand the basis for cleft lip and palate in both animals and man.

References

1. Clegg, D. J. (1971). Teratology. *Annu. Rev. Pharmacol.*, **11**, 409
2. Jost, A. and Picon, L. (1970). Hormonal control of fetal development and metabolism. *Adv. Metabol. Disorders*, **4**, 123
3. Rutter, W. J., Pictet, R. L. and Morris, P. L. (1973). Towards molecular mechanisms of developmental processes. *Annu. Rev. Biochem.*, **42**, 601
4. Ballard, P. L. and Tomkins, G. M. (1970). Glucocorticoid-induced alteration of the surface membrane of cultured hepatoma cells. *J. Cell Biol.*, **47**, 222
5. Cake, M. H. and Litwack, G. (1975). The glucocorticoid receptor. In G. Litwack (ed.). *Biochemical Actions of Hormones*. Vol. 3, pp. 317–340. (New York: Academic Press)
6. Solomon, S. and Lee, D. K. M. (1977). Binding of glucocorticoids in fetal tissues. *J. Steroid Biochem.*, **8**, 453
7. Salomon, D. S., Zubairi, Y. and Thompson, E. B. (1978). Ontogeny and biochemical properties of glucocorticoid receptors in midgestation mouse embryos. *J. Steroid Biochem.*, **9**, 95
8. Kalter, H. (1954). Inheritance of susceptibility to the teratogenic action of cortisone in mice. *Genetics*, **39**, 185
9. Baxter, H. and Fraser, F. C. (1950). Production of congenital defects in the offspring of female mice treated with cortisone. *McGill Med. J.*, **19**, 245
10. Dostal, M. and Jelinek, R. (1973). Sensitivity of embryos and interspecies differences in mice in response to prenatal administration of corticoids. *Teratology*, **8**, 245
11. Marsk, L., Running, K. and Larsson, K. S. (1974). ^{3}H-cortisone incorporation in CBA and transferred A/J embryos in CBA mothers. *Biol. Neonate*, **24**, 49
12. Walker, B. E. (1967). Induction of cleft palate in rabbits by several glucocorticoids. *Proc. Soc. Exp. Biol. Med.*, **125**, 1281

13. Shah, R. N. and Kilistoff, A. (1976). Cleft palate induction in hamster fetuses by glucocorticoid hormones and their synthetic analogues. *J. Embryol. Exp. Morphol.*, **36**, 101
14. Zarrow, M. X., Philpott, J. E. and Denenberg, V. H. (1970). Passage of ^{14}C-corticosterone from the rat mother to fetus and neonate. *Nature*, **226**, 1058
15. Nguyen-Trong-Tuan, Rekdal, D. J. and Burton, A. F. (1971). Uptake and metabolism of ^3H-cortisone and fluorimetric determination of corticosterone in fetuses of several mouse strains. *Biol. Neonate*, **18**, 78
16. Badr, F. M. and Spickett, S. G. (1965). Genetic variation in the biosynthesis of corticosteroids in *Mus musculus*. *Nature*, **205**, 1088
17. Doering, C. H., Shire, J. G. M., Kessler, S. and Clayton, R. B. (1972). Cholesterol ester concentrations and corticosterone production in adrenals of the C57BL/10 and DBA/2 strains in relation to adrenal lipid depletion. *Endocrinology*, **90**, 93
18. Lindberg, M., Shire, J. G. M., Doering, C. H., Kessler, S. and Clayton, R. B. (1972). Reductive metabolism of corticosterone in mice: Differences in NADPH requirements of liver homogenates of males of two inbred strains. *Endocrinology*, **90**, 81
19. Zimmerman, E. F. and Bowen, D. (1972). Distribution and metabolism of triamcinilone acetonide in inbred mice with different cleft palate sensitivities. *Teratology*, **5**, 335
20. Levine, A. I., Yaffe, S. J. and Back, N. (1968). Maternal-fetal distribution of radioactive cortisol and its correlation with teratogenic effect. *Proc. Soc. Exp. Biol. Med.*, **129**, 86
21. Reminga, T. A. and Avery, J. K. (1972). Differential binding of labeled cortisone in fetal, placental and maternal liver tissue in A/J mice. *J. Dent. Res.*, **51**, 1426
22. Wood, N. K., Murks, A. D., Schmitz, D. C., Bowman, D. C. and Toto, P. D. (1972). Radioautographic labeling in placenta and fetal palatal shelves after maternal injection of ^3H-cortisone in the A/J mouse. *J. Dent. Res.*, **51**, 67
23. Spain, K. M., Disieleski, W. and Wood, N. K. (1975). Cleft palate induction: Quantitative studies of ^3H-corticoids in A/J mouse after maternal injections of ^3H-cortisol. *J. Dent. Res.*, **54**, 1069
24. Salomon, D. S., Gift, V. D. and Pratt, R. M. (1979). Corticosterone levels during midgestation in the maternal plasma and fetus of cleft palate-sensitive and -resistant mice. *Endocrinology*, **104**, 154
25. Barlow, S. M., Morrison, P. J. and Sullivan, F. M. (1974). Plasma corticosterone levels during pregnancy in the mouse: The relative contributions of the adrenal glands and fetoplacental units. *J. Endocrinol.*, **60**, 473
26. Barlow, S. M., Morrison, P. J. and Sullivan, F. M. (1975). Effects of acute and chronic stress on plasma corticosterone levels in pregnant and nonpregnant mice. *J. Endocrinol.*, **66**, 93
27. Sugimoto, M., Kojima, A. and Endo, H. (1976). Role of glucocorticoids in terminal differentiation and tissue-specific function. *Dev. Growth Differ.*, **18**, 319
28. Salomon, D. S. and Pratt, R. M. (1976). Glucocorticoid receptors in murine embryonic facial mesenchyme cells. *Nature*, **264**, 174
29. Salomon, D. S. and Pratt, R. M. (1978). Inhibition of growth in vitro by glucocorticoids in mouse embryonic facial mesenchyme cells. *J. Cell. Physiol.*, **97**, 315
30. Goldman, A. S., Katsumata, M., Yaffe, S. J. and Gassee, D. L. (1977). Palatal cytosol cortisol-binding protein associated with cleft palate susceptibility and H-2 genotype. *Nature*, **265**, 643
31. Yamamoto, K. R., Gehring, M., Stampfer, M. and Sibley, C. H. (1976). Genetic approaches to steroid hormone action. In R. O. Greep (ed.). *Recent Progress in Hormone Research*, Vol. 32, pp. 3–52. (New York: Academic Press)
32. Hackney, J. F., Gross, S. R., Aronow, L. and Pratt, W. B. (1970). Specific glucocorticoid-binding macromolecules from mouse fibroblasts in vitro. *Mol. Pharmacol.*, **6**, 500
33. Nanda, R. and Romeo, D. (1978). Effect of dexamethasone and vitamin A on cellular proliferation of rat palatal processes. *Cleft Palate J.*, **15**, 176
34. Nanda, R., Van der Linden, P. F. G. M. and Jansen, H. W. B. (1970). Production of cleft palate with dexamethasone and hypervitaminosis A in rat embryos. *Experientia*, **26**, 1111
35. Mott, W. J., Toto, P. D. and Hilgers, D. C. (1969). Labelling index and cellular density in palatine shelves of cleft palate mice. *J. Dent. Res.*, **48**, 263
36. Jelinek, R. and Dostal, M. (1975). Inhibitory effects of corticoids on the proliferative pattern in the mouse palatal processes. *Teratology*, **11**, 193

37. Baker, J. B., Barsh, G. S., Carney, D. H. and Cunningham, D. D. (1978). Dexamethasone modulates binding and action of epidermal growth factor in serum-free cell culture. *Proc. Natl. Acad. Sci. (USA)*, 75, 1882
38. Gospodarowicz, D. and Moran, J. S. (1976). Growth factors in mammalian cell culture. *Annu. Rev. Biochem.*, 45, 531
39. Nacht, S. and Garzon, P. (1974). Effects of corticosteroids on connective tissue and fibroblasts. In: M. H. Briggs and C. A. Christie (eds.). *Advances In Steroid Biochemistry And Pharmacology*, Vol. 4, pp. 157–187. (New York: Academic Press)
40. Moscatelli, D. and Rubin, H. (1977). Hormonal control of hyaluronic acid production in fibroblasts and its relation to nucleic acid and protein synthesis. *J. Cell. Physiol.*, 91, 79
41. Shapira, Y. and Shoshun, S. (1972). The effect of cortisone on collagen synthesis in the secondary palate of mice. *Arch. Oral Biol.*, 17, 1699
42. Ballard, P. L., Baxter, J. D., Higgins, S. J., Rousseau, G. G. and Tomkins, G. M. (1974). General presence of glucocorticoid receptors in mammalian tissues. *Endocrinology*, 94, 998
43. Berliner, J. A. and Gerschenson, L. E. (1975). The effects of a glucocorticoid on the cell surface of RLC-CAI cells. *J. Cell. Physiol.*, 86, 523
44. Walker, B. E. and Fraser, F. C. (1957). The embryology of cortisone-induced cleft palate. *J. Embryol. Exp. Morphol.*, 5, 201
45. Pratt, R. M., Goggins, J. R., Wilk, A. L. and King, C. T. G. (1973). Acid mucopolysaccharide synthesis in the secondary palate of the rat at the time of rotation and fusion. *Dev. Biol.*, 32, 230
46. Toole, B., Jackson, G. and Gross, J. (1972). Hyaluronate in morphogenesis: Inhibition of chondrogenesis in vitro. *Proc. Natl. Acad. Sci. (USA)*, 69, 1384
47. Wilk, A. L., King, C. T. G. and Pratt, R. M. (1978). Chlorcyclizine induction of cleft palate in the rat: Degradation of palatal glycosaminoglycans. *Teratology*, 18, 199
48. Pratt, R. M. and King, C. T. G. (1971). Collagen synthesis in the secondary palate of the developing rat. *Arch. Oral Biol.*, 16, 1181
49. Hassell, J. R. and Orkin, R. W. (1976). Synthesis and distribution of collagen in the developing palate. *Dev. Biol.*, 53, 86
50. Pratt, R. M. and King, C. T. G. (1972). Inhibition of collagen cross-linking associated with BAPN-induced rat cleft palate. *Dev. Biol.*, 27, 322
51. Greene, R. M. and Pratt, R. M. (1976). Developmental aspects of secondary palate formation. *J. Embryol. Exp. Morphol.*, 36, 225
52. Hayward, F. (1969). Ultrastructural changes in the epithelium during fusion of the palatal processes in rats. *Arch. Oral Biol.*, 14, 661
53. Pratt, R. M. and Hassell, J. R. (1975). Appearance and distribution of carbohydrate-rich macromolecules on the epithelial surface of the developing rat palate. *Dev. Biol.*, 45, 192
54. Greene, R. M. and Kochhar, D. M. (1974). Surface coat on the epithelium of developing palatine shelves in the mouse. *J. Embryol. Exp. Morphol.*, 31, 683
55. Greene, R. M. and Pratt, R. M. (1977). Inhibition by DON of rat palatal glycoprotein synthesis and epithelial cell adhesion in vitro. *Exp. Cell Res.*, 105, 27
56. Yamada, K. M., Yamada, S. S. and Pastan, I. (1975). The major cell surface glycoprotein of chick embryo fibroblasts is an agglutinin. *Proc. Natl. Acad. Sci. (USA)*, 72, 3158
57. Morgan, P. R. (1976). The fate of the expected fusion zone in rat fetuses with experimentally induced cleft palate. An EM study. *Dev. Biol.*, 51, 225
58. Hudson, C. D. and Shapiro, B. L. (1973). A radioautographic study of DNA synthesis in the embryonic rat palatal shelf epithelium. *Arch. Oral Biol.*, 18, 72
59. Pratt, R. M. and Martin, G. R. (1975). Epithelial cell death and cyclic AMP increase during palatal development. *Proc. Natl. Acad. Sci. (USA)*, 72, 874
60. Mato, M., Aikana, E. and Katahira, M. (1966). Appearance of various types of lysosomes in the epithelium covering lateral palatine shelves during secondary palate formation. *Gumma J. Med.*, 15, 46
61. Lorente *et al.* (1974). Lysosomal enzyme associated with palatal development in the rabbit. *J. Dent. Res.*, 53, 65
62. Smiley, G. R. (1970). Fine structure of mouse embryonal palatal epithelium prior to and after midline fusion. *Arch. Oral Biol.*, 15, 287

63. Smiley, G. R. and Koch, W. E. (1972). An in vitro and vivo study of single palatal processes. *Anat. Rec.*, 173, 405
64. Tyler, M. S. and Koch, W. E. (1975). In vitro development of palatal tissue from embryonic mice. I. Differentiation of the secondary palate from day 12 mice. *Anat. Rec.*, 182, 297
65. Tyler, M. S. and Koch, W. E. (1977). In vitro development of palatal tissues from embryonic mice. II and III. *J. Embryol. Exp. Morphol.*, 38, 19
66. Pratt, R. M. and Greene, R. M. (1976). Inhibition of palatal epithelial cell death in vitro by altered protein synthesis. *Dev. Biol.*, 54, 125
67. Greene, R. M. and Pratt, R. M. (1979). Inhibition of epithelial cell death in the secondary palate in vitro by altered lysosome function. *J. Histochem. Cytochem.*, 26, 1109
68. Greene, R. M. and Pratt, R. M. (1979). Correlation between localization of adenylate cyclase and transient increase in cyclic AMP in the rodent palate in vivo. *J. Histochem. Cytochem.* (In press)
69. Pastan, I., Johnson, G. S. and Anderson, W. B. (1975). Role of cyclic nucleotides in growth control. *Annu. Rev. Biochem.*, 44, 491
70. Waterman, R. E., Parker, G. C., Palmer, S. J. and Palmer, S. M. (1976). Catecholamine-sensitive adenylate cyclase in the developing golden hamster. *Anat. Rec.*, 185, 125
71. Cohen, S. (1962). Precocious eyelid opening and tooth eruption in the rat by EGF. *J. Biol. Chem.*, 237, 1555
72. Hassell, J. R. (1975). An ultrastructural study of the inhibition of epithelial cell death and palatal fusion by EGF. *Dev. Biol.*, 45, 90
73. Hassell, J. R. and Pratt, R. M. (1977). Elevated levels of cAMP alters the effect of EGF in vitro on programmed cell death in the secondary palate epithelium. *Exp. Cell. Res.*, 106, 55
74. Smuts, M. S. (1977). CON A binding to the epithelial surface of the developing mouse olfactory placode. *Anat. Rec.*, 188, 29
75. Wilson, D. B. and Hendricks, A. G. (1977). Quantitative aspects of proliferation in the nasal epithelium of the Rhesus monkey embryo. *J. Embryol. Exp. Morphol.*, 38, 217
76. Johnston, M. C. and Pratt, R. M. (1974). The neural crest in normal and abnormal craniofacial development. In H. C. Slavkin and R. C. Greulich (eds.). *Extracellular Matrix Influences on Gene Expression*, pp. 773–777. (New York: Academic Press)
77. Minkoff, R. and Kuntz, A. J. (1977). Cell proliferation during morphogenetic change: Analysis of frontonasal morphogenesis in the chick embryo employing DNA labeling indices. *J. Embryol. Exp. Morphol.*, 40, 101
78. Trasler, D. G. (1968). Pathogenesis of cleft lip and its relation to embryonic face shape in A/J and C57 mice. *Teratology*, 1, 33
79. Trasler, D. G. and Fraser, F. C. (1977). Time-position relationships. In J. G. Wilson and F. C. Fraser (eds.). *Handbook of Teratology*, pp. 271–292. (New York: Plenum Press)
80. Lejour, M. (1970). Cleft lip induced in the rat. *Cleft Palate J.*, 7, 169
81. Trasler, D. G. and Leong, S. (1974). Face shape and mitotic index in mice with 6AN-induced and inherited cleft lip. *Teratology*, 9, 39A
82. Johnston, M. C., Sulik, K. K. and Dudley, K. H. (1978). Phenytoin-induced cleft lip and palate in mice. *Teratology*, 17, 29A
83. Figueroa, A. A. and Pratt, R. M. (1979). Autoradiographic study of macromolecular synthesis in the fusion epithelium of the developing rat primary palate in vitro. *J. Embryol. Exp. Morphol.*, 50, 145
84. Gaare, J. D. and Langman, J. (1977). Fusion of nasal swellings in the mouse embryo: Regression of the nasal fin. *Am. J. Anat.*, 150, 461
85. Figueroa, A. A. and Greene, R. M. (1977). Cyclic AMP levels in the developing mouse primary palate. *Teratology*, 17, 30A

11
In vitro studies of virus-induced disorders of prenatal growth and development

A. D. HEGGIE

INTRODUCTION

Certain maternal viral infections during pregnancy have been identified as causes of death or abnormal development in the mammalian embryo and fetus[1,2]. Although the number of viruses clearly identified as being embryopathic is small, it serves to demonstrate the potential of viral infections as inducers of reproductive failure and emphasizes the need for an understanding of the interactions that may take place between viruses and the developing organism. Most investigations of these interactions have been conducted *in vivo*. Pregnant animals have been inoculated with viruses and then the effects of maternal viral infection on the outcome of pregnancy have been observed[3,4]. *In vivo* studies are essential for investigation of many aspects of gestational infection and much has been learned from them. They often fail, however, in differentiating direct embryopathic effects of viral infection from indirect effects related to disordered physiology in the infected dame and they cannot provide for direct and continuous observation of the developing embryo or fetus. Evaluation of viral effects in studies conducted *in vivo* is limited to examination of pre-term fetuses after sacrifice of the pregnant animal or of the young after birth. By these techniques events such as arrest of cleavage, failure of implantation, or death of the embryo during the pre-implantation stage are difficult to detect. *In vitro* studies, in contrast, permit direct and continuing observation of the response of embryonic cells, fetal organs, or whole embryos to viral exposure under controlled conditions. Although it must be emphasized that these controlled conditions are artificial and that the response of embryos in this milieu cannot be assumed to be equivalent to what occurs *in vivo*, investigations conducted *in vitro* can make unique and important contributions to an understanding of the pathogenesis of virus-induced disorders of development when their results are interpreted in conjunction with observations made *in vivo*. The purpose of this review is to substantiate the utility of the *in vitro* approach.

THE RUBELLA SYNDROME
Background

The rubella syndrome is probably the best example of a virus-induced disorder of human development. Affected infants often have congenital cataracts, cardiovascular anomalies, nerve deafness, microcephaly, and a variety of other abnormalities including size and weight which are disproportionately small for gestational age[5]. No animal model for reproducing this combination of defects is available. Investigation of the pathogenesis of the rubella syndrome, therefore, has depended upon *in vivo* and post-mortem examination of infected infants and *in vitro* observations of the reaction of human embryonic cells and organs to infection with rubella virus. Because of the limitations of the studies which can be performed *in vivo* in affected infants, the contributions of *in vitro* studies have been important.

Effects on mitosis

By infecting monolayer cultures of human diploid lung fibroblasts with rubella virus, Plotkin *et al.* noted that infected cells gradually lost their capacity to replicate, although they appeared morphologically indistinguishable from uninfected fibroblasts[6]. A similar effect was produced in cell cultures derived from lung, pituitary, kidney, and thymus obtained from 8–16-week human embryos. Like infection of embryos with rubella virus *in utero*, infection of embryonic cells *in vitro* was persistent. This impaired multiplication of cells chronically infected with rubella virus was confirmed and quantified by Rawls and Melnick[7]. They prepared cell cultures from infected organs of infants who died from the complications of congenital rubella and compared the time required for these cultures to double their cell number with that of cells derived from uninfected organs. Figure 11.1 shows the results of these experiments. Suspensions of cells were seeded into culture vessels and after an initial decrease in cell number, probably related to the stress of transfer, cells adhered to the vessel wall and began to multiply. Uninfected control cells then doubled their number in 24 h. This generation time was obtained repeatedly on subsequent passages of control cells. Cells infected with rubella virus, however, were found to have a doubling time of 48 h on early subcultivations which increased to 72 h after 10 or more passages. Although it cannot be concluded solely from the results of these *in vitro* experiments that impaired cell multiplication occurs in the infected embryo *in vivo*, these findings are compatible with observations such as the small size of many infants with congenital rubella and the decreased number of cells which their body organs contain[8] and they suggest a mechanism by which these abnormalities can be reasonably explained.

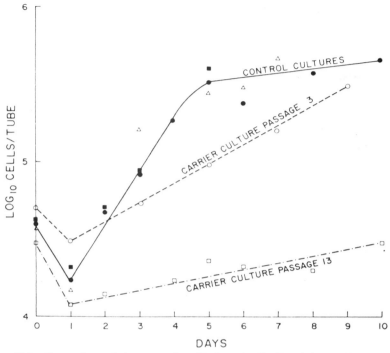

Figure 11.1 Comparison of the rates of replication of cells from infants with congenital rubella in uninfected control cultures and in rubella-infected carrier cultures. From Rawls, W. E. and Melnick, J. L. (1966). *J. Exp. Med.*, 123, 795[7]

Retarded bone growth

Supportive evidence that mitotic inhibition may contribute to the retardation of growth in infants with congenital rubella was provided by the studies of Heggie showing that the growth of human embryonic long bones in organ culture was impaired by inoculation with rubella virus[9]. Growth curves of bones in cultures inoculated with live and inactivated rubella virus are presented in Figure 11.2. Growth, as measured by increase in the mean wet weights of bones, was approximately 40% less in cultures inoculated with live virus than in cultures inoculated with inactivated virus. This finding was of additional interest in view of the transient radiolucent lesions observed in the long bones of many infants with congenital rubella[10,11]. These lesions are thought to be the result of a delay in growth of bony trabeculae[12].

Chromosome breaks

Another mechanism suggested by *in vitro* studies as a contributing cause of retarded growth in congenital rubella is chromosome breakage. In their initial studies of the *in vitro* growth of human embryonic cells infected with rubella virus, Plotkin *et al.* found that 50% of mitoses in infected cells had chromo-

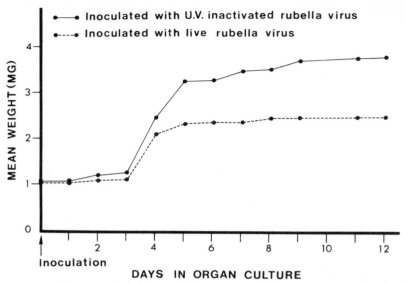

Figure 11.2 Growth inhibition of 8-week human embryonic long bones in organ cultures inoculated with live rubella virus. Bones in cultures inoculated with ultraviolet inactivated virus show normal growth. From Heggie, A. D. (1977). *Teratology*, 15, 47[9]

some breaks or gaps, while these changes were seen in only 10% of uninfected cells[6]. Further studies of cells derived from embryos infected with rubella virus *in utero*, however, showed that the increase in number of chromosome breaks in infected cells, although demonstrable, was small[13]. Boué and Boué continued this line of investigation *in vitro* by determining the number of

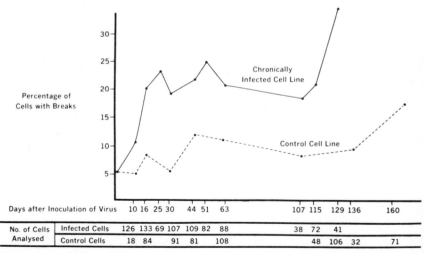

Figure 11.3 Increased frequency of chromosome breaks in human diploid cell cultures chronically infected with rubella virus. From Boué, A. and Boué, J. G. (1969). *Am. J. Dis. Child.*, **118**, 45 (Copyright 1969, American Medical Association)[14]

chromosome breaks in cultures of human embryonic cells inoculated with rubella virus. As shown in Figure 11.3, the percentage of chromosome breaks in infected cells was found to be at least twice that which occurred in uninfected control cells[14]. It was postulated that, if chromosome breaks occur during cell division, the daughter cells which incorporate the defective chromosomes will not be viable and, therefore, that an increased rate of chromosome breaks in the cells of rubella infected embryos may account for their retarded growth rate *in vivo*.

Cataract formation

The mechanism of cataract formation in congenital rubella has been explored in an elegant study by Karkinen–Jääskeläinen *et al.*[15]. Eye rudiments from human embryos were grown as organ cultures. The effects of exposure to

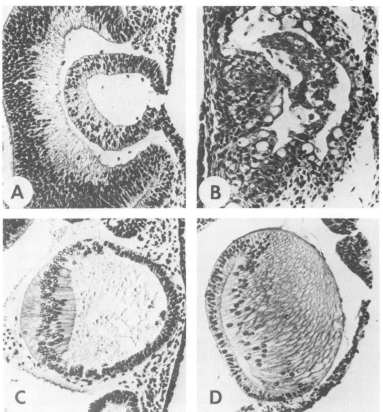

Figure 11.4 Rubella cataract formation in organ cultures of human eye rudiments. A, Open lens vesicle of a 4–5 week embryo at time of rubella virus inoculation. B, Vacuolar degeneration of lens fibres two weeks after inoculation with virus in the open-vesicle stage. C, Closed lens vesicle stage of a 6-week embryo at time of virus inoculation. D, Normal differentiation of lens two weeks after inoculation with virus in the closed vesicle stage. From Karkinen-Jääskeläinen, M., Saxén, L., Vaheri, A. and Leinikki, P. (1975). *J. Exp. Med.*, **141**, 1238[15]

rubella virus on lens development *in vitro* were compared in rudiments from 4–6 week embryos in the open lens vesicle stage and from 6–10 week embryos in which the lens vesicles had closed. Lens development in uninfected control rudiments was observed concurrently. Uninfected rudiments in the open lens vesicle stage developed normally, showing invagination of the lens placode, closure of the lens vesicle and normal differentiation of lens fibres. Rudiments infected in the open vesicle stage developed normally for about 10 days when in contrast to controls, vacuolar degeneration of differentiating lens fibres began (Figure 11.4). Progression of these degenerative changes in the lens resulted in a histological appearance similar to that seen in the cataracts of congenital rubella. In eye rudiments infected after closure of the lens vesicle, development was similar to that in uninfected controls (Figure 11.4). If, however, a small incision was made in the capsule of the closed lens vesicle prior to exposure to rubella virus, degeneration of lens fibres ensued in a manner similar to that in rudiments infected before closure of the vesicle. From these data it was concluded that the lens capsule becomes an impermeable barrier to rubella virus at the time of closure of the lens vesicle and thereafter protects against cataract formation by preventing rubella virus from reaching the sensitive lens fibres. If virus has infected the fibres before capsule formation, however, degeneration and cataract formation will occur and a high concentration of virus may accumulate within the confines of the virus-impermeable lens capsule. These concepts are consistent with the observations *in vivo* that cataracts in congenital rubella occur in association with maternal rubella early in the first trimester[16] and that cataractous lenses in rubella affected infants contain higher concentrations of rubella virus than other tissues[17], and may remain infected long after virus has been eliminated from all other organs[18].

DIFFERENTIATION AND SUSCEPTIBILITY TO VIRUS INFECTION

Exposure of embryonic organ cultures to viruses has been used successfully to assess changes in the susceptibility of various tissues to infection at different stages of development. Saxén *et al.*[19] and Vainio *et al.*[20], for example, demonstrated with cultures of 12–16 day mouse kidney rudiments that undifferentiated metanephrogenic mesenchymal cells were readily infected and destroyed by exposure to polyoma virus. As soon as mesenchymal cells began to undergo differentiation into convoluted renal tubules, however, they became resistant to polyoma virus infection and to its cytolytic effect (Figure 11.5). In contrast to these findings *in vitro*, newborn mice, including their kidneys, are known to be susceptible *in vivo* to infection and to tumour induction by polyoma virus. This discrepancy between *in vitro* and *in vivo* responses does not invalidate the results of studies conducted *in vitro*, but rather suggests the need to search for modifying factors which may be operative in the intact animal.

IN VITRO STUDIES IN MATERNAL VIRAL INFECTION

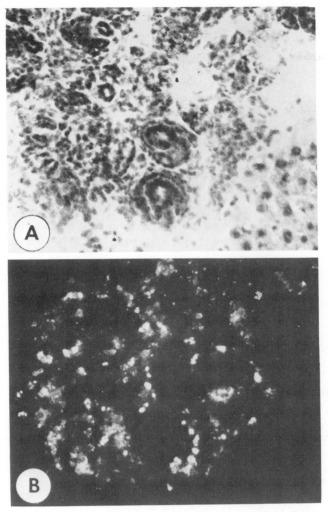

Figure 11.5 Susceptibility of undifferentiated mesenchymal cells and resistance of tubular cells to polyoma virus infection in a kidney rudiment from a 16-day mouse embryo. A, Photomicrograph stained with haematoxylin and eosin showing newly formed secretory tubules surrounded by undifferentiated mesenchymal cells. B, Same section stained with fluorescein-labelled mouse anti-polyoma virus immune globulin showing presence of fluorescence corresponding to viral antigen in mesenchymal cells and its absence in secretory tubules. From Vainio, T., Saxén, L. and Toivonen, S. (1963). *Virology*, 20, 380[20]

H-1 VIRUS INFECTION OF HAMSTER TIBIAE
Background

Infection with the H-1 strain of parvovirus is teratogenic and highly embryocidal in the hamster[3, 21, 22]. Inoculation of newborn animals with this virus produces a syndrome which has been compared to Down's syndrome in humans[23]. No chromosomal abnormalities appear to be involved but these animals are dwarfed and show craniofacial disproportion, defective teeth and decreased learning ability. *In vivo* investigations strongly suggest that the defects observed in teeth and bones of affected animals are the direct result of infection of osseous tissues by H-1 virus. The possibility had to be considered, however, that these effects on bone were attributable to indirect mechanisms, since retarded growth, especially in the very immature animal, is often a complication of chronic infection. Also, since H-1 virus infects many tissues, it was possible that apparent isolation of virus from bone *in vivo* might represent contamination of bone by infected adjacent tissues or blood.

Retarded growth of infected tibiae

To limit the variables encountered *in vivo* and study the effects of virus exposure on bone in a controlled environment, Heggie exposed organ cultures of 13–15 day hamster embryonic tibiae to H-1 virus[24]. Virus was found to replicate readily in bone and as shown in Figure 11.6 to inhibit bone growth.

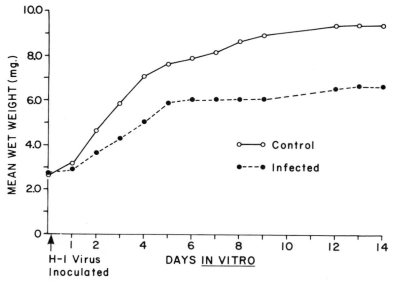

Figure 11.6 Growth inhibition in 15-day embryonic hamster tibiae infected with H-1 virus in organ culture. From Heggie, A. D. (1971). *J. Exp. Med.*, 133, 506[24]

Histologically, infected bones showed periosteal and perichondral degeneration and diminished deposits of subperiosteal bone. Thus it was concluded that embryonic hamster bones are susceptible to infection with H-1 virus and that the infection has a direct effect on bone growth.

Altered mucopolysaccharide metabolism

In an attempt to identify a biochemical lesion by which H-1 virus infection induces growth retardation in these bones, Heggie and Stjernholm compared the rates of mucopolysaccharide synthesis and cartilage formation in infected and uninfected tibiae in this organ culture model[25]. S^{35} uptake was used as a measurement of mucopolysaccharide synthesis and mineral-free dry weight as an indicator of the cartilage content of the bones. Mean S^{35} uptake was found to be three times greater in infected than in uninfected tibiae. Release of S^{35} from bone after transfer to isotope-free medium was also greater in infected than in control tibiae. This was felt to reflect an increase in the turnover (production and loss) of mucopolysaccharide in infected tibiae. Mineral-free dry weight, which was assumed to be a function primarily of the amount of cartilage present in embryonic tibiae, was 21% less in infected than in control bones. Differences in both S^{35} uptake and mineral-free dry weight were statistically significant ($p < 0.05$ by the matched pair t-test). As a result of these findings it was proposed that H-1 virus infection interferes with cartilage formation and that, as has been observed in osteoarthritic cartilage[26], increased polysaccharide synthesis reflects a compensatory but ineffective effort to reverse this inhibition. Although there was no appreciable difference between the mineral contents (ash weights) of infected and control tibiae, it was suggested that defective cartilage formation may contribute to the delayed or incomplete mineralization of osseous tissues that occurs *in vivo* during later stages of maturation in this animal model of virus-induced dwarfism.

VIRUSES AND PREIMPLANTATION EMBRYOS
Background

Death of the embryo in the preimplantation stage is usually an undetected event *in vivo*. Investigations of the effects of viral infection on the preimplantation embryo, therefore, are best conducted *in vitro*. By exposing two-cell embryos to the mengo strain of encephalomyocarditis virus *in vitro*, Gwatkin was first to demonstrate that a virus could infect and replicate in the preimplantation embryo and that this infection resulted in arrest of cleavage and degeneration of blastomeres[27-30]. In similar tests with other viruses, including herpes simplex[31], simian virus-40[32-34], Moloney sarcoma virus[32,33], polyoma virus[34] and minute virus of mice[35], various investigators were unable to detect any interference with cleavage or blastocyst formation.

Degeneration of preimplantation embryos after exposure to coxsackievirus B-4

In a recent report Heggie described degeneration of preimplantation mouse embryos after exposure to coxsackievirus B-4 *in vitro*[36]. In these experiments embryos were harvested at the two-cell stage and placed in cultures which were then immediately inoculated with virus. Subsequent cleavage and blastocyst formation in embryos in virus-inoculated cultures were compared with those in embryos in cultures inoculated with control fluids. More than 60% of embryos in cultures inoculated with control fluids progressed to the blastocyst stage. In embryos exposed to coxsackievirus B-4, however, cleavage was arrested at the eight-cell stage and degeneration of blastomeres ensued (Table 11.1 and Figure 11.7). An embryo so affected would almost certainly not be

Table 11.1 Effect of viral exposure of mouse embryos at the two-cell stage on subsequent blastocyst formation *in vitro*

Virus tested	Size of virus inoculum	Uninoculated control	Inoculated with control fluid	Virus inoculated
Coxsackie type B-4	10^5 TCID$_{50}$	149/163 (91)*	118/178 (66)†	16/210 (8)‡
Coxsackie type B-6	10^4 TCID$_{50}$	33/44 (75)	57/94 (61)	15/95 (16)‡
ECHO type 11	10^5 TCID$_{50}$	147/181 (81)	152/169 (90)	150/168 (89)
Attenuated polio type 2	10^5 TCID$_{50}$	81/94 (86)	72/83 (87)	82/91 (90)
Reovirus type 2	10^3 TCID$_{50}$	69/96 (72)	83/108 (77)	45/108 (42)‡
Influenza type A	64 HA units	72/78 (92)	96/110 (87)	78/106 (74)‡
Parainfluenza type 1	16 HA units	54/61 (89)	44/69 (64)†	55/93 (59)
Mumps	32 HA units	70/75 (93)	47/71 (66)†	51/80 (64)
Rubella	10^4 TCID$_{50}$	98/113 (87)	110/161 (68)†	119/178 (68)
Herpes simplex type 1	10^4 TCID$_{50}$	87/104 (84)	79/114 (69)†	64/101 (63)
Herpes simplex type 2	10^3 TCID$_{50}$	91/109 (84)	87/113 (77)	77/115 (67)
Mouse cytomegalovirus	10^3 TCID$_{50}$	70/75 (93)	47/71 (66)†	19/44 (43)‡
Adenovirus type 5	10^5 TCID$_{50}$	46/51 (90)	59/93 (63)†	47/99 (48)‡
Mouse adenovirus	10^4 TCID$_{50}$	57/64 (89)	64/80 (80)	12/86 (14)‡

* No. of embryos forming blastocysts/total no. of embryos in culture (%)
† Statistically significant difference in blastocyst yield between uninoculated control cultures and cultures inoculated with control fluid ($p < 0.05$ by the chi-square test)
‡ Statistically significant difference in blastocyst yield between cultures inoculated with control fluid and cultures inoculated with virus ($p < 0.05$ by the chi-square test)
From Heggie, A. D. (1979). *Pediatr. Res.* (In press)[36]

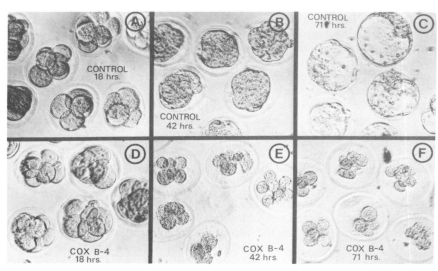

Figure 11.7 Cleavage and blastocyst formation in cultures of two-cell mouse embryos after exposure to coxsackievirus B-4 or control fluids *in vitro*. A–C, Normal progression through 8-cell, morula, and blastocyst stages at 18, 42, and 71 h after inoculation with control fluid. D, Progression to 8-cell stage at 18 h after inoculation with coxsackievirus B-4; but, E, F, arrest of cleavage and degeneration of blastomeres at 42 and 71 h after viral inoculation. From Heggie, A. D. (1979). *Pediatr. Res.* (In press)[36]

viable *in vivo*. To determine that these effects were related to live virus and not to some product of viral infection, embryos were exposed to ultraviolet inactivated virus. Exposure to inactivated virus did not produce degeneration or arrest of cleavage, indicating that these effects required the presence of live virus. The embryopathic potential of the coxsackie group of viruses, therefore, makes them likely causes of reproductive failure if contact with the preimplantation embryo occurs. This contact apparently does not occur in the mouse, since studies by Soike[37] showed that inoculation of female mice with coxsackievirus B-3 just before mating had no effect on the outcome of pregnancy. Maternal infection induced later in pregnancy, however, resulted in transplacental infection and fetal wastage. Whether or not the genital tract of the human female is permeable to coxsackieviruses during early pregnancy is not known.

Inhibition of blastocyst formation by other viruses

Two-cell mouse embryos were exposed to a spectrum of other viruses (Table 11.1) and several were found to inhibit blastocyst formation. The rapid degeneration of blastomeres that was associated with coxsackievirus B-4 was not observed, however. A point to be emphasized is that the development of preimplantation mouse embryos *in vitro* is easily altered by a variety of environmental factors. Each set of experiments summarized in Table 11.1 included not only virus-inoculated and uninoculated cultures but also cultures inoculated with control fluids which consisted of medium and the products of disrupted cells harvested from the same batch of cultures or eggs used for preparation of the virus stock being tested. Thus control fluids were identical to virus stocks except that they did not contain virus. As shown in Table 11.1, inoculation of embryo cultures with control fluids for seven of the 14 viruses tested resulted in statistically significant reductions in blastocyst yields when compared with uninoculated cultures. These reductions were probably caused by inhibitory products of cell metabolism present in the control fluids and presumably, therefore, in the virus stocks. This type of control must be included to distinguish reductions in blastocyst yields caused by non-viral inhibitors from reductions caused by the interaction of viruses with embryos.

Failure of a cytocidal virus to interfere with blastocyst formation

Exposure of two-cell mouse embryos to herpes simplex viruses types 1 and 2 (HSV-1 and HSV-2) did not interfere with blastocyst formation (Table 11.1) or result in discernable embryopathic changes. This was surprising in view of the lethal infection that is produced by these viruses in suckling and adult mice. Therefore, to assess the virulence of the strains of HSV being used and to ensure that embryos were in constant contact with the virus, two-cell embryos were co-cultivated with monolayer cultures of primary African green monkey kidney cells that were then infected immediately with HSV-1 or HSV-2. Although cytopathic effects characteristic of HSV-1 or HSV-2 promptly appeared in the monkey kidney cell monolayers, no cytopathic changes were noted in the embryos resting on these monolayers and there

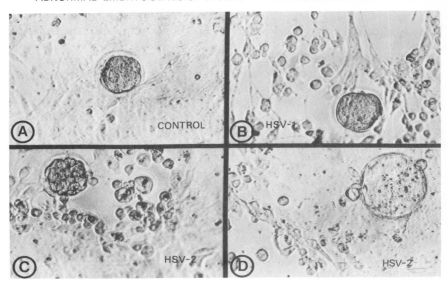

Figure 11.8 Mouse embryos co-cultivated from the two-cell stage with monolayer cultures of African green monkey kidney cells inoculated with herpes simplex viruses types 1 or 2 or with control fluid. A, Normal morula resting on cell monolayer 44 h after inoculation with control fluids. B, C, Normal morulae on cell monolayers showing viral cytopathic effects and degeneration 44 h after inoculation with herpes simplex viruses type 1 and type 2 respectively. D, Normal blastocyst on cell monolayer showing viral simplex virus type 2. From Heggie, A. D. (1979). *Pediatr. Res.* (In press)[36]

was no inhibition of blastocyst formation (Figure 11.8). These observations indicate that the response of the preimplantation embryo to viral exposure cannot be reliably predicted from the effects produced by exposure of the adult or suckling host to the same virus. It must also be recognized that in these investigations embryonic development was observed only through the blastocyst stage. Therefore, virus-induced defects may have been present, but inapparent at this early developmental stage. Experiments involving transfer of virus-exposed, but apparently normal, blastocysts to the uteri of pseudopregnant animals and observation of the outcomes of these pregnancies must be conducted before it can be concluded that no virus-induced abnormalities occurred as a result of virus exposure of the embryo during the preimplantation stage.

EMBRYOS AND ONCOGENIC VIRUSES
Simian virus-40

Although the primary consideration of this review to this point has been the role of viruses as causes of abnormal development, there are important similarities between the developmental effects of viruses and their oncogenic potential. Oncogenesis, teratogenesis and reproductive failure are all forms of growth dysfunction. What are the effects of exposing early embryos to oncogenic viruses *in vitro*? Exposure of preimplantation mouse embryos to simian virus-40 (SV-40), polyoma and Moloney sarcoma viruses has been found to result in infection of the embryos without causing degeneration of blastomeres or interfering with cleavage and blastocyst formation *in vitro*[32-34]. Infection with SV-40 and Moloney sarcoma virus required prior removal of the zona pellucida which is apparently impermeable to these viruses[32]. Jaenisch and Mintz intensified exposure of early embryos by injecting SV-40 viral DNA into the blastocoeles of explanted mouse blastocysts which were then transferred to the uteri of pseudopregnant surrogate mothers[38]. Development was then allowed to proceed to term *in utero* and the offspring were studied. Although there were no controls with uninoculated or sham-inoculated blastocysts, the survival rate to birth and the long-term survival rate of animals derived from blastocysts injected with SV-40 DNA were comparable to those of uninfected blastocysts in other embryo transfer studies conducted in the laboratory of these investigators. DNA, extracted from the tissues of long-term surviving offspring of surrogate mothers, was tested for the presence of SV-40 specific sequences. These sequences were detected in at least one organ of 40% of the animals tested and this finding was felt to be consistent with the hypothesis that SV-40 DNA was integrated into the host genome during development. There was no evidence of expression of viral genetic functions such as tumor formation during observation of these animals to one year of age.

Moloney murine leukaemia virus

To continue studies with oncongenic viruses, Jaenisch *et al.* exposed zona-free mouse embryos at the 4–8 cell stage to Moloney murine leukaemia virus (MMLV) *in vitro*[39]. This was done by adding virus to the culture fluids rather than by injection of the blastocoele. After reaching the blastocyst stage *in vitro*, embryos were transferred to pseudopregnant recipients to complete development. Of the offspring of these surrogate mothers, MMLV viraemia was detected in approximately 7% by four to eight weeks of age. In tests on one of these viraemic animals, a male, slow annealing MMLV-specific DNA sequences were identified in all organs, including the testis. These sequences were not present in the organs of uninfected control mice. In mice that developed leukaemia as a result of inoculation with MMLV after birth, rather than at the 4–8 cell stage, these MMLV–DNA sequences were detectable only in 'target' organs infiltrated by leukaemia cells. They were not found in the uninvolved organs of these animals. These data were felt to be compatible with the concept that exposure of the preimplantation embryo to MMLV resulted in integration of virus into all cells of the animal and that in certain animals leukaemia occurred as a result of expression of the MMLV viral genome. Viraemia was detected in 5% of the offspring born from matings of normal females with the viraemic male originally exposed to MMLV as a 4–8 cell embryo. Matings of this male's viraemic 'sons' (N-1 generation) and 'grandsons' (N-2 generation) with normal females, however, produced offspring in which the frequency of viraemia was approximately 50%[40,41]. MMLV–DNA sequences were detected in all organs of the latter viraemic mice. Matings of normal females with viraemic males infected with

MMLV as newborns produced non-viraemic offspring. These data were regarded as evidence that exposure of preimplantation mouse embryos to MMLV results in integration of virus into the germ cells of some animals and that passage of viral genetic information to subsequent generations is by Mendelian transmission.

POSSIBILITIES FOR ADDITIONAL INVESTIGATIONS

Many other *in vitro* models of developmental processes are available. They include experimental systems for *in vitro* observation of the aggregation of dissociated embryonic cells[42,43], attachment of blastocysts and outgrowth of trophoblast cells[44], formation of egg cylinders and development to the stage of early somites[45-47], organogenesis, and growth and development of post-implantation embryos[48]. Exposure of these models to the effects of viruses offers an opportunity to further evaluate these infectious agents as potential causes of abnormal development.

ACKNOWLEDGEMENT

The author's investigative work discussed in this review was supported by National Institutes of Health grant HD-04110, U.S. Public Health Service.

References

1. Hardy, J. B. (1973). Fetal consequences of maternal viral infection in pregnancy. *Arch. Otolaryngol.*, 98, 218
2. Dudgeon, J. A. (1976). Infective causes of human malformations. *Br. Med. Bull.*, 32, 77
3. Kilham, L. and Margolis, G. (1975). Problems of human concern arising from animal models of intrauterine and neonatal infections due to viruses: a review. I. Introduction and virologic studies. II. Pathologic studies. *Progr. Med. Virol.*, 20, 114
4. Elizan, T. S. and Fabiyi, A. (1970). Congenital and neonatal anomalies linked with viral infections in experimental animals. *Am. J. Obstet. Gynecol.*, 106, 147
5. Forrest, J. M. and Menser, M. A. (1975). Recent implications of intrauterine and postnatal rubella. *Aust. Paediatr. J.*, 11, 65
6. Plotkin, S. A., Boué, A. and Boué, J. G. (1965). The in vitro growth of rubella virus in human embryo cells. *Am. J. Epidemiol.*, 81, 71
7. Rawls, W. E. and Melnick, J. L. (1966). Rubella virus carrier cultures derived from congenitally infected infants. *J. Exp. Med.*, 123, 795
8. Naeye, R. L. and Blanc, W. (1965). Pathogenesis of congenital rubella. *J. Am. Med. Assoc.*, 194, 1277
9. Heggie, A. D. (1977). Growth inhibition of human embryonic and fetal rat bones in organ culture by rubella virus. *Teratology*, 15, 47
10. Rudolph, A. J., Singleton, E. B., Rosenberg, H. S., Singer, D. B. and Phillips, C. A. (1965). Osseous manifestations of the congenital rubella syndrome. *Am. J. Dis. Child.*, 110, 428
11. Singleton, E. B., Rudolph, A. J., Rosenberg, H. S. and Singer, D. B. (1966). The roentgenographic manifestations of the rubella syndrome in newborn infants. *Am. J. Roentgen.*, 97, 82
12. Reed, G. B. (1969). Rubella bone lesions. *J. Pediatr.*, 74, 208
13. Chang, T. H., Moorehead, P. S., Boué, J. G., Plotkin, S. A. and Hoskins, J. M. (1966). Chromosome studies of human cells infected 'in utero' and 'in vitro' with rubella virus. *Proc. Soc. Exp. Biol. Med.*, 122, 236
14. Boué, A. and Boué, J. G. (1969). Effects of rubella virus infection on the division of human cells. *Am. J. Dis. Child.*, 118, 45

15. Karkinen-Jääskeläinen, M., Saxén, L., Vaheri, A. and Leinikki, P. (1975). Rubella cataract in vitro: sensitive period of the developing human lens. *J. Exp. Med.,* 141, 1238
16. Murphy, A. M., Reid, R. R., Pollard, I., Gillespie, A. M., Dorman, D. C., Menser, M. A., Harley, J. D. and Hertzberg, M. B. (1967). Rubella cataracts. Further clinical and virologic observations. *Am. J. Ophthalmol.,* 64, 1109
17. Bellanti, J. A., Artenstein, M. S., Olson, L. C., Buescher, E. L., Luhrs, C. E. and Milstead, K. L. (1965). Congenital rubella. Clinicopathologic, virologic, and immunologic studies. *Am. J. Dis. Child.,* 110, 464
18. Menser, M. A., Harley, J. D., Hertzberg, R., Dorman, D. C. and Murphy, A. M. (1967). Persistence of virus in lens for three years after prenatal rubella. *Lancet,* ii, 387
19. Saxén, L., Vainio, T. and Toivonen, S. (1962). Effect of polyoma virus on mouse kidney rudiment in vitro. *J. Natl. Can. Inst.,* 29, 597
20. Vainio, T., Saxén, L. and Toivonen, S. (1963). The acquisition of cellular resistance to polyoma virus during embryonic differentiation. *Virology,* 20, 380
21. Ferm, V. H. and Kilham, L. (1965). Histopathologic basis of the teratogenic effects of H-1 virus on hamster embryos. *J. Embryol. Exp. Morphol.,* 13, 151
22. Ferm, V. H. and Kilham, L. (1965). Skeletal studies of virus-induced dwarfism. *Growth,* 29, 7
23. Toolan, H. W. (1960). Experimental production of mongoloid hamsters. *Science,* 131, 1446
24. Heggie, A. D. (1971). Pathogenesis of H-1 virus infection of embryonic hamster bone in organ culture. *J. Exp. Med.,* 133, 506
25. Heggie, A. D. and Stjernholm, R. L. (1973). Altered mucopolysaccharide metabolism in organ cultures of fetal hamster bones infected by the H-1 strain of parvovirus. *Teratology,* 8, 147
26. McElligott, T. F. and Collins, D. H. (1960). Chondrocyte function of human articular and costal cartilage compared by measuring the in vitro uptake of labelled (^{35}S) sulphate. *Ann. Rheumatic Dis.,* 19, 31
27. Gwatkin, R. B. L. (1963). Effect of viruses on early mammalian development. I. Action of mengo encephalitis virus on mouse ova cultivated in vitro. *Proc. Natl. Acad. Sci. (USA),* 50, 576
28. Gwatkin, R. B. L. (1964). Effect of viruses on early mammalian development. II. Synthesis of meningoencephalitis virus by mouse ova maintained in vitro. *J. Cell. Boll.,* 23, 40A
29. Gwatkin, R. B. L. (1966). Effect of viruses on early mammalian development. III. Further studies concerning the interaction of meningoencephalitis virus with mouse ova. *Fertil. Steril.,* 17, 411
30. Gwatkin, R. B. L. and Auerbach, S. (1966). Synthesis of a ribonucleic acid virus by the mammalian ovum. *Nature (Lond.),* 209, 993
31. Berkovich, S. (1963). An attempt to infect with virus the early rabbit embryo maintained in vitro. *Abstracts of the 33rd Annual Meeting of the Society for Pediatric Research,* p. 81
32. Sawicki, W., Baranska, W. and Koprowski, H. (1971). Susceptibility of unfertilized and fertilized mouse eggs to simian virus 40 and Moloney sarcoma virus. *J. Natl. Cancer Inst.,* 47, 1045
33. Baranska, W., Sawicki, W. and Koprowski, H. (1971). Infection of mammalian unfertilized and fertilized ova with oncogenic viruses. *Nature (Lond.),* 230, 591
34. Biczysko, W., Solter, D., Pienkowski, M. and Koprowski, H. (1973). Interactions of early mouse embryos with oncogenic viruses – simian virus 40 and polyoma. I. Ultrastructural studies. *J. Natl. Cancer Inst.,* 51, 1945
35. Mohanty, S. B. and Bachmann, P. A. (1974). Susceptibility of fertilized mouse eggs to minute virus of mice. *Infect. Immun.,* 9, 762
36. Heggie, A. D. (1979). Effects of viral exposure of the two-cell mouse embryo on cleavage and blastocyst formation in vitro. *Pediatr. Res.* (In press)
37. Soike, K. (1967). Coxsackie B-3 virus infection in the pregnant mouse. *J. Infect. Dis.,* 117, 203
38. Jaenisch, R. and Mintz, B. (1974). Simian virus 40 DNA sequences in DNA of healthy adult mice derived from preimplantation blastocysts injected with viral DNA. *Proc. Natl. Acad. Sci. (USA),* 71, 1250
39. Jaenisch, R., Fan, H. and Croker, B. (1975). Infection of preimplantation mouse embryos

and of newborn mice with leukemia virus: tissue distribution of viral DNA and RNA and leukemogenesis in the adult animal. *Proc. Natl. Acad. Sci. (USA)*, 72, 4008

40. Jaenisch, R. (1976). Germ line integration and Mendelian transmission of the exogenous Moloney leukemia virus. *Proc. Natl. Acad. Sci. (USA)*, 73, 1260
41. Jaenisch, R., Berns, A., Dausman, J. and Cox, V. (1976). Germ line integration and leukemogenesis of exogenous and endogenous murine leukemia viruses. In: D. Baltimore, A. S. Huang and C. F. Fox (eds.). *Animal Virology. ICN–UCLA Symposium on Molecular and Cellular Biology*. Vol. IV, pp. 283–310. (New York: Academic Press)
42. Moscona, A. A. (1961). Rotation mediated histogenetic interaction of dissociated cells. *Exp. Cell Res.*, 22, 455
43. Moscona, A. A. (1965). Recombination of dissociated cells and the development of cell aggregates. In: E. N. Willmer (ed.). *Cells and Tissues in Culture*, pp. 489–529. (New York: Academic Press)
44. Sherman, M. I. (1975). Long term culture of cells derived from mouse blastocysts. *Differentiation*, 3, 51
45. Hsu, Y.-C. (1973). Differentiation in vitro of mouse embryos to the stage of early somite. *Dev. Biol.*, 33, 403
46. Pienkowski, M., Solter, D. and Koprowski, H. (1974). Early mouse embryos: growth and differentiation in vitro. *Exp. Cell Res.*, 85, 424
47. Juurlink, B. H. J. and Federoff, S. (1977). Effects of culture milieus on the development of mouse blastocysts in vitro. *In Vitro*, 13, 790
48. New, D. A. T. (1971). Methods for the culture of post-implantation embryos of rodents. In: J. C. Daniel (ed.). *Methods in Mammalian Embryology*, pp. 305–319. (San Francisco: Freeman)

12
Impaired adaptation to extrauterine life: a teratogenic event

R. DE MEYER AND M. PEETERS

INTRODUCTION

The field of teratology has long been restricted to the morphological aspects of developmental anomalies. Since the major morphogenic processes are ended as early as the first trimester of gestation, the late gestational period has been considered to be free of teratogenic risk. However, it has been observed that interference during late pregnancy could result in fetal growth retardation and late fetal as well as neonatal mortality.

Both have long been thought to be caused by the toxicity of interfering factors rather than to teratogenic processes. In some cases this may be true, in so far as well-established physiological and biochemical functions may be disturbed or that destructive lesions occur in some parenchyma. For example, if a drug induces necrosis of liver cells or destroys red blood cells, then one must consider this as a toxic and not as a teratogenic effect. In other cases, however, growth retardation and fetal mortality could be the consequence of a faulty development of vital functions or metabolic processes; for instance if one imagines the lack of maturation of a particular enzyme or metabolic pathway.

One would be tempted to call this kind of teratogenic process 'metabolic'. This is not entirely justified, since even morphological teratogenesis is the consequence of a metabolic disturbance that is usually not well understood. For example, the abnormal limb development caused by thalidomide must be mediated by the disturbance of some metabolic pathway, although this pathway remains unknown. Developmental anomalies with no morphological expression should therefore be referred to as 'functional' teratology.

The term teratology is extensive and covers a wide range of anomalies. For instance:

 1. A function vital to the fetus in late pregnancy may be disturbed and

result in fetal death. It can be difficult to distinguish whether this is due to a simple toxic effect or to some teratological event.

2. Anomalies may occur in the perinatal period due to altered development of the functions that are necessary for the adaptation to extrauterine life.

3. Since development is not terminated at birth, many anomalies can be induced during the postnatal period. Because birth occurs at various stages of development depending on the species, the same metabolic event can occur before or after birth. It is consequently important that the postnatal period be considered also in the evaluation of teratogenic risk.

The present work will cover only anomalies that occur in the perinatal period of the rat and that are due to the altered development of the functions which are necessary for the adaptation to extrauterine life. The perinatal period is the most appropriate one for studying developmental anomalies which arose during intrauterine life. Indeed, it is at birth that the neonate has to suddenly cope with many new functions. It is, therefore, only at birth, when these functions must become operative, that anomalies, having occurred at various stages of the prenatal development, are expressed.

The early neonatal period is, in this respect, a very good one for testing the antenatal development of many functions. In this chapter on adaptation to extrauterine life, three main points will be examined:

1. Some metabolic modifications (both antenatal and conatal) that occur during normal adaptation.

2. Intrauterine growth retardation, prematurity and postmaturity as disturbances in the adaptation to extrauterine life.

3. Chemical factors with an influence on the developmental stages that permit normal adaptation.

SOME FUNCTIONAL MODIFICATIONS OCCURRING DURING THE PERINATAL PERIOD: A STUDY OF THEIR NATURE

The aim of this chapter is to describe some of the functional modifications which occur during the perinatal period. Then the mechanisms that trigger these changes will be specified. The neonatal period is characterized by:

1. A change in metabolism which shifts from anabolic to catabolic due to the interruption of maternal nutritive supply; and

2. The neonate assuming functions previously performed by the mother (respiration, digestion and detoxication). In order to cope with these requirements, the newborn must mobilize its energy reserves (carbohydrates, proteins and lipids), and initiate its pulmonary and gastrointestinal functions.

The following will be successively analysed:

(a) The changes in carbohydrate metabolism
(b) The changes in protein metabolism
(c) The changes in lipid metabolism
(d) Hormonal changes and their function as factors of adaptation.

Changes in carbohydrate metabolism

Modifications in metabolite concentration

(a) *Glucose*. Fetal blood glucose is always less than to that found in the mother[1-4], although it increases during late gestation[2,4,5]. It is generally admitted that there is a transient phase of hypoglycaemia after birth[1,4,6-12], the importance of which is rated differently. The importance of this drop depends on the temperature at which neonates are kept: the closer the temperature is to 37 °C, the more profound is the hypoglycaemia[13,14].

(b) *Lactic acid*. Towards the end of gestation, the total plasma concentration of lactate is approximately 100 mg/dl. Within the 60 to 90 minutes after birth, the concentration of lactic acid becomes normal and reaches values close to those found in the adult (± 10 mg/dl). This drop is due to the fact that the rate at which it is used exceeds its rate of production[5,6,15,16].

(c) Blood pyruvate concentration was found to rise from 0.12 μmol/l at birth to 0.16 μmol/l within one hour after birth, at which value it remains stable[1].

(d) A study made on the concentration of citrate, one of the metabolites of the Krebs cycle, showed that its concentration fell from 8 mg/dl to 4 mg/dl in two hours[17].

(e) *Glycogen*. Many analyses of rat liver glycogen[1,4,-7,10,18,19] have indicated that its concentration at birth is high. The glycogen reserves can be mobilized, but only after some delay[6,9,13,20]. The amount of glycogen found in the carcass has also been measured[5,6]. It accounts for approximately 50% of the total amount of glycogen in the animal[5,20]. This reserve, which is largely used during the first six hours and too often neglected in the evaluation of energy metabolism, must be taken into account.

Modifications of enzyme activity

(a) *Enzymes of glycolysis*. Very few modifications in the concentration of these enzymes have been detected during the perinatal period[21-25]. This can be readily understood if one keeps in mind the importance of this metabolic pathway which is essential for antenatal metabolism. Metabolic flow through the glycolytic pathway is three times less than in the adult (0.63 μmol/min/g)[26].

Two enzymes deserve special attention: glucokinase and fructose-1,6-diphosphatase. They are not essential for glycolysis, although both are branched on to the pathway. The action of the first one (glucokinase) can be replaced by hexokinase, the rate limiting enzyme which is already present at a very early stage in the embryo[27,28]. Glucokinase, on the other hand, differentiates only after the fifteenth postnatal day[29-31]. The second enzyme – fructose-1,6-diphosphatase – is active mainly in the conversion of fructose diphosphate to fructose monophosphate, i.e. in the direction of the anabolic pathway. The early presence of the latter enzyme, which constitutes an important branching and control point, allows the glycolytic pathway to be functional during the first phases of embryogenesis[32]. As far as fructose-1,6-diphosphatase is concerned, its enzyme activity starts just before birth, on the twenty-first day[32-34].

Table 12.1 summarizes the developmental changes of some enzymes intervening in glycolysis[23,27-34]. It is noteworthy that a change in enzymatic

Table 12.1 Changes during development of some glycolytic enzymes (activity expressed as % of adult values)

Enzyme	Gestational age (days)		Birth		Days after birth		Adult	Referer
	19th	21st	0'	12 h	5th	20th		
Hexokinase	130	130			90	85	100	29
Total activity	400	325	310	300	250	150	100	28
Isozyme activity:								
A	250	300	290	280	200	100	100	
B	200	260	300	360	220	85	100	27
C	75	90	100	120	260	100	100	
Glucokinase	0	0	0	0	0	25	100	30, 31
	0	0	0	0	2	20	100	29
Fructose-1,6-diphosphatase	58	87	87	90	107	138	100	32
	22	22	66	89	156	246	100	34
Fructokinase	266	146	133		67	100*	—	23
	275	220	200	180	150	150	100	28

* Percentage calculated for 100% on the twentieth day

activity is not an exclusive aspect of differentiation; changes in kinetics have also been observed and are related to the appearance of particular isozymes[27].

(b) *Enzymes of the Krebs cycle.* Table 12.2 shows the differentiation of the principal enzymes or groups of enzymes involved in the Krebs cycle[32, 35, 36]. The low rate of activity of aconitase makes it the rate-limiting enzyme of this cycle. The slow increase in the activity of this enzyme from the moment of birth is a partial explanation of how the Krebs cycle starts functioning (before birth the activity of the Krebs cycle is very low)[24, 26, 37]. This low activity could

Table 12.2 Changes during development of some enzymes of the Krebs cycle (expressed as % of adult values)

Enzyme	Before birth	After birth (days)			Reference
		1	2	10	
Citrate synthetase	95	95	95	95	35
Aconitase:					
Mitochondria	10	30	40	100	35
Cytoplasma	15	15			35
Isocitrate dehydrogenase	65 100	98	/	98	32
α ketoglutarate dehydrogenase	95	95	95	95	35
Succinate thiokinase	/	/	/	/	
Succinic dehydrogenase	60	75	80	100	36
Fumarase	50	50	50	90	35
Malate dehydrogenase	72	83		94	32
NAD specific	98	90		93	36
NADP specific	5	5		5	32

be related to the small number of mitochondria in the cell[38], as well as to the low concentration of enzymes in the mitochondria[36].

(c) *Enzymes involved in the synthesis and in the catabolism of glycogen.* Many enzymes intervene in this metabolic pathway. For some of these we have but limited data concerning their role in the antenatal and perinatal development. It has been shown that *in utero* only the anabolic pathway is used, although the enzymes for glycogenolysis are also present[39]. The moment of significant increase in the rate of activity of different enzymes is shown in Table 12.3[19, 33, 39–46]. The two key enzymes are glycogen synthetase and phosphorylase. In their case, activity is not solely due to their presence, since these enzymes can exist under both active and inactive forms. The interconversion between these two forms is controlled by kinases.

Synthetase is present as early as the sixteenth gestational day[46] but its activation is only possible around the eighteenth day, when the activity of synthetase phosphatase suddenly increases[39]. It has been found that the activity of this enzyme corresponds to the onset of glycogen storage occurring at the same moment[39, 46]. As far as the breakdown of glycogen is concerned, there exists under normal circumstances no glycogenolysis *in utero*, although

Table 12.3 Moment of appearance of significant enzyme activity

Enzyme	Moment of appearance of significant enzyme activity (gestational day)	Reference
Phosphoglucomutase	18–19	42
	15–18	19
	18	43
UDPG pyrophosphorylase	18	41
	16–18	19
Glycogen synthetase	18	19
	18	39
	18–19	42
	19	43
	15.5–18	46
Branching enzyme (amylo 1,4-1,6 transglucosylase)	/	
Debranching enzyme (amylo 1,6 glucosidase)	/	
Phosphorylase (total)	19–20	42
	19	39
	18–19*	19
	19	45
	19	43
	19	46
Glucose-6-phosphatase	19–20†	43
	20	45
	20†	33
	20–21	42

* Second peak at the moment of birth
† Second peak after birth

phosphorylase is found before birth, which would indicate that the enzyme is present mostly in its inactive non-phosphorylated form. This unfortunately has not yet been demonstrated, due to technical difficulties[39]. Indeed, the conversion of phosphorylase from its inactive to its active form occurs very rapidly when the animals are killed. The activity of protein kinase and phosphorylase kinase has not been measured accurately during the neonatal period. However, these enzymes seem to be present before birth, which allows glycogenolysis *in utero* under particular circumstances such as anaesthesia[39].

Modifications in protein metabolism

Modifications in metabolite concentration

(a) Plasma amino acid concentration is found to be, on the whole, markedly higher in the neonate than in the mother[47]. However, a sharp drop is observed within the first hours[47].

(b) Blood plasma urea is low in the fetus and the excretion of urea during the first two weeks after birth is also minimal[48–50]. A study of urea formation from liver slices indicated that the capacity to synthesize urea increases very significantly during the first day of extrauterine life[50].

The modification of some enzymes intervening in protein metabolism

(a) The gluconeogenesis pathway is inoperative *in utero* and becomes functional during the first day of extrauterine life[6, 47, 51–53] (essentially during the first six hours[47] and possibly even during the first hour[6]).

Various enzymes have been studied *in vitro*; the most noteworthy of which are pyruvate carboxylase and phosphoenolpyruvate carboxykinase (PEPCK)[54]. Pyruvate carboxylase can be detected from the eighteenth gestational day and increases until it reaches at birth one-third of the adult activity[55, 56]. The concentration of PEPCK is very low before birth; therefore it is the rate-limiting enzyme[8, 33, 57–59]. However, its sudden increase just after birth enables the activation of the gluconeogenic pathway. *In vitro*, the liver acquires the ability to transform radioactive pyruvate into glucose from the moment PEPCK appears[60]. It has been proved both *in vivo* and *in vitro* that the gluconeogenic pathway becomes opeational shortly after birth.

(b) Of the five enzymes intervening in the urea cycle, only one (arginine synthetase) has an activity level which is almost nil until birth when it exhibits a sudden increase in enzyme activity[50, 51, 61–63]. Before birth, the four other enzymes of this cycle have a capacity which is between 10 and 20% of the values in adults[50]. Their values also increase during the postnatal period, but more slowly than that observed in the case of arginine synthetase.

(c) Tyrosine aminotransferase is an enzyme favouring the catabolism of amino acids. It allows the transfer of ammonia from tyrosine to alpha ketoglutarate with the formation of glutamate. Glutamate can regenerate alpha ketoglutarate liberating ammonia, which can enter the urea cycle. The tyrosine, thus liberated, can be degraded to hydroxyphenylpyruvic acid which will then be catabolized further. The enzyme activity of tyrosine amino transferase increases suddenly just before birth[4, 64, 65]. Tyrosine amino transferase is thus a witness of the metabolic flow towards the amino acid catabolic pathway.

Changes in lipid metabolism

The newborn rat differs from neonates of other animal species in that it has no white adipose tissue[1]. The triglyceride reserves found in brown adipose tissue and in the liver are insignificant[1]. The quantity of fat found in the newborn of different animal species is shown in Table 12.4[66-68]. This insufficiency has

Table 12.4 Fat content in newly born mammals

Mammal	Fat (g/kg)	Reference
Human:		
Term	161	
Premature	50	66
Rat	11	
Lamb	30	67
Pig	11	
Rabbit	58	68

important metabolic implications. Aside from brown adipose tissue, which plays an important role in the thermoregulation of the newborn and which will not be dicussed further here, fatty acid oxidation plays only a small role as energy source during the perinatal period.

The absence of white adipose tissue can be accounted for in two ways, i.e. the relative impermeability of the placenta to fatty acids[69, 70], and the absence of certain enzymes intervening in fatty acid metabolism which develop only after birth. As far as substrates derived from fat are concerned, it seems that ketone bodies are a relatively more important source of energy during the neonatal period than they are later on[71].

Modifications in metabolite concentration

(a) *Free fatty acids*. Free fatty acid plasma concentrations found in the newborn are significantly lower than those found in the mother. During the first six hours after birth there is a 50% increase in the values observed[1].

(b) *Glycerol*. Blood glycerol, which is low at birth, rises during the first hour after birth[1]. Subsequently it returns to its initial value within six hours.

(c) *Beta hydroxybutyrate*. Plasma concentration of beta hydroxybutyrate is higher in the fetus than in the mother. There is a rapid drop after birth and the values observed remain low until the sixth postnatal hour[55, 72].

Modifications of enzyme activity

The enzymes intervening in fatty acid synthesis can be classified as follows: enzymes found in the cytoplasm intervening in the synthesis of short fatty acid chains from acetyl-CoA; and enzymes located in the mitochondria and in microsomes and intervening in the fatty acid chain elongation.

It has been shown that in the fetal rat liver the activity of fatty acid synthetase is higher than that of the fatty acid elongation system[73-75]. The oxidation of fatty acid chains occurs in the mitochondria of extrahepatic tissues.

The acetyl-CoA formed as product of the fatty acid oxidation system may then enter the tricarboxylic acid cycle.

An important step in fatty acid oxidation is the entry of fatty acids into mitochondria from the cytoplasm. The transfer of the fatty acyl group from CoA to the carrier molecule carnitine is due to the action of an enzyme fatty acyl CoA – carnitine fatty acid transferase.

Little is known about the perinatal development of enzymes intervening in the oxidation and in the biosynthesis of lipids. As far as acetyl CoA carboxylase is concerned, there is a transient peak in its enzyme activity at the moment of birth. This momentary increase is followed by a decrease in enzyme activity[73]. It has been shown that the carnitine acetyl transferase activity in the rat liver increases slowly during the perinatal period and reaches its peak around the fourth postnatal day[76].

The capacity for long chain fatty acid oxidation in the liver, which is low during fetal life, increases sharply after birth and remains high until weaning[77, 78]. The perinatal development of the enzyme activity of beta hydroxybutyrate dehydrogenase has been studied. Its activity is almost nil until the twentieth gestational day. There is a sudden and rapid increase in its activity starting on the twenty-first day[35]. It must be underlined that this enzyme is also found in the brain, where it catalyses the oxidation of beta hydroxybutyrate to acetoacetate, which can be further metabolized[77, 79, 80]. Although the brain does not play a direct role in the adaptation to extrauterine life, the fact that it can derive energy from ketone bodies shows that these are an alternative energy source and this has a sparing effect on carbohydrates.

Hormonal changes and their function as factors of adaptation

Numerous endocrine factors undergo a change during the neonatal period, the survey of which is still incomplete. The present discussion will be limited to the description of certain changes which are well-established and significant.

Plasma corticosterone concentrations

These increase during normal vaginal delivery, but are progressively restored to normal during the 12 h that follow[81, 82]. In the case of Caesarean section, concentration is normal at birth but rises during the two hours that follow[83], at which time it reaches the level observed in rats born by vaginal delivery[84]. This difference makes it possible to distinguish two types of stress: that occurring at birth and that occurring during the postnatal period.

The secretion of insulin and glucagon

This has been much discussed[1, 2, 10, 11, 85–88]. All investigators agree that the plasma insulin concentration, which is high during late gestation, decreases before delivery and falls even lower during the six hours following delivery. Glucagon, on the contrary, rises during the first half-hour following delivery, then falls slightly until the beginning of the sixth hour, but it still maintains a value higher that that observed before birth.

Although catecholamines have been considered as an important factor in the homeostasis of glucose in the adult, they do not play a major role in the newborn[14]. This is mainly attributed to a lack of response of the adrenal gland to the usual stimuli rather than to a secretory incompetence. Thus, for instance, hypoglycaemia and hypothermia seem inactive, whereas hypoxia is a very good stimulus[14].

Thyroid function

A follicular structure is noted in the fetal rat thyroid gland on the eighteenth gestational day[89]. At this time it develops the capacity to uptake and accumulate iodide[90]. It was thought that the rat placenta was impermeable for thyroxine[91]. Recently it has been shown that, although the placenta is readily permeable to thyroxine[92], there is evidence that towards the end of gestation there is increased secretion of this hormone by the fetal thyroid[90,92]. However, there seems to be little or no placental transfer of T_3 in the rat[93]. The fetal and neonatal plasma concentrations of thyroxine and of tri-iodothyronine remain low until the fifth postnatal day, after which there is an abrupt increase in the production rate. Maximal values occur around the sixteenth postnatal day[94,95].

The enzyme activity of monoamine oxidase in the thyroid gland has been shown to increase significantly after birth. Adult values are reached on the sixteenth postnatal day[96]. Deiodinase was found to have low enzyme activity at birth whereas enzyme activity of tyrosine aminotransaminase was high[97]. Serial measurements of serum and pituitary TSH, of hypothalamic TRH and of serum T_3 and T_4 have shown that, during the postnatal period, the components of the hypothalamic–pituitary–thyroid axis develop simultaneously[94].

It seems that immediately at birth there is a brief discharge of pituitary TSH and of the thyroid hormone stores[98]. It has been shown *in vitro* that there exists a pituitary responsiveness to TRH before birth[99]. Nevertheless, rats are hypothyroid at birth. However, it must be noted that in this animal species the processes usually considered as dependent on thyroid hormones develop only after birth (for example skeletal and cerebellar maturation).

The relation between the hormonal changes and the metabolic modifications occurring during the early postnatal period can by summarized as follows:

1. Changes in enzyme activity could be explained by an increase in glucagon and a decrease in insulin levels during the early neonatal period. Indeed glucagon has a stimulating effect on glucose-6-phosphatase[1,45,100], on PEPCK[1,45,59,101], on phosphorylase[1,45,59] and also on the incorporation of amino acids in the liver[102], while glycogen synthetase is inhibited[13]. On the other hand, insulin has an inhibitory effect on PEPCK, does not affect the uptake of amino acids by the liver[102], but does stimulate their uptake by muscle tissue[103,104]. A combined action of increasing glucagon and decreasing insulin levels would explain the onset of glycogenolysis and postnatal gluconeogenesis. Glucagon action has been thought to occur through the adenylate cyclase–cyclic AMP system[1,8,12,105–107], but now doubt has been cast on this hypothesis[108]. Indeed, it has been shown that there is a relative membrane receptor insufficiency[108].

2. It has been said that the changes in glucagon and insulin secretions described above are secondary to hypoglycaemia[58,110,111]; this, however, has been questioned[13]. Therefore, the role of catecholamines and of the nervous

system has been postulated as a regulatory mechanism. The presence of nerve receptors in the pancreas is an important argument in favour of this hypothesis[14]. It has also been shown that the secretion of glucagon and insulin is modified after the administration of adrenalin[13]. However, the medulla of the adrenal gland seems not to respond to hypoglycaemia and only hypoxia appears to be an adequate stimulus[14].

This leads one to suppose that hypoxia is the *primum movens* of the metabolic changes in the neonatal period. It would be worthwhile to investigate the role of other parameters, such as pCO_2 and lactacidaemia, which also undergo great changes just after birth.

Conclusion

Metabolic adjustment to extrauterine life is a complex phenomenon. Two types of processes must be distinguished:

1. Adaptative processes that occur before birth but that will only be expressed after birth. For example, the phosphorylase which the newborn needs at birth for the glycogenolysis can be found before birth but remains in its inactive form.

2. Adaptative changes which are induced by the birth process itself. They are of two types:

(a) Activation of an enzyme which, as mentioned earlier, developed before birth but remained inactive; for example, the activation of phosphorylase after birth allowing glycogenolysis. This activation process is important and compulsory. Indeed, if the induction of an enzyme occurs *in utero*, but the function of this enzyme appears only after birth, its activation must be delayed until after birth. It must, therefore, be maintained in its inactive form until the moment when the metabolic process for which it is indispensable must be initiated.

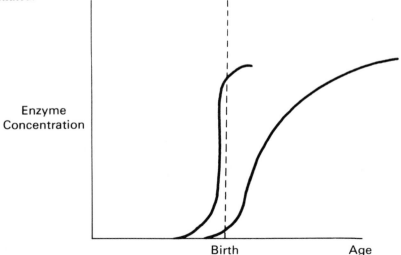

Figure 12.1 Postnatal induction of enzymes

(b) *Postnatal induction of enzymes*. An example of this type is the increase of PEPCK and the initiation of the gluconeogenic pathway, or the induction of arginine synthetase which permits the onset of the urea cycle.

The two types of processes are illustrated in Figure 12.1[25].

INTRAUTERINE GROWTH RETARDATION, PREMATURITY AND POSTMATURITY AS TERATOGENIC FACTORS

Intrauterine growth retardation can be produced in different ways:

(a) A reduction of uterine blood flow by means of ligation of the afferent uterine vessels[112]
(b) A restriction of maternal nutrition[113]
(c) The use of hydroxyurea[114]
(d) The use of antiblastic drugs[115].

The newborns

The newborns affected by intrauterine growth retardation demonstrate the following characteristics: a weight reduction of approximately 30%[116,117]; and a body composition which is characterized by the following:

(a) A modification of the weight distribution of various organs: the liver accounts for only 4.42% of the total wet body weight instead of the 6.34% found in controls, and the brain accounts for 4.90% instead of 3.59%[112,116].

(b) A modification of the global chemical composition. The water content is found to be increased slightly[116], the concentration of plasma proteins is lower than that found in controls, whereas the protein content of the organs is normal[117]. Since the concentration of DNA in the liver is normal, this might be interpreted to mean that the total number of cells per gram of hepatic tissue is normal. However, it seems that the ratio of the different types of cells (hepatocytes, erythrocytes and erythropoietic cells) could possibly be altered.

Modifications in metabolites, enzymes, hormones and metabolism

Metabolites

As far as blood glucose is concerned, its value is lower at birth and after birth a greater fall is observed[15,117-120]. Lactate concentration is found to be normal[121]. Pyruvate values have not been determined. Hepatic glycogen concentration and its mobilization are normal at birth[118,120,121]. Since the liver is smaller in the fetus suffering from intrauterine growth retardation, the total quantity of glycogen available is reduced (5.08 mg versus 17.92 mg)[116]. Total plasma amino acid levels were found to be normal or higher than in controls[122].

Enzymes

There is but little data available concerning the development of enzymes in rats with intrauterine growth retardation. It has been shown that glucose-6-phosphatase and fructose-1,6-diphosphatase differentiate in these animals, just as they do in the controls[15]. Phosphoenolpyruvate carboxykinase

(PEPCK) has been shown to develop normally[123]. However, some authors have suggested that there is a delay in the induction of this enzyme[51]. The moment of the onset of gluconeogenesis is normal[123].

Hormones

Although the insulin concentration at birth was found to be lower in rats with intrauterine growth retardation than in the controls, the postnatal evolution was identical in both groups[88, 123].

Conversely, the levels of glucagon at birth were higher in the experimental animals but fell rapidly to normal soon afterwards[88, 123]. There are no data available concerning the modifications of the levels of adrenalin and corticoids which might exist. It would be important to obtain such information, given the decisive role that these two hormones play in the normal process of adaptation to extrauterine life.

Metabolism

There is a profound hypoglycaemia at the moment of birth which persists after birth[15, 117-120]. The antenatal administration of corticoids has no effect on the hypoglycaemia which occurs at birth, but prevents the occurrence of a secondary one[120]. As a consequence of these experiments, it is possible to make a distinction between the hypoglycaemia occurring at birth and that of the postnatal period. It is, therefore, likely that each of these episodes has its own determining factors. Postnatal hypoglycaemia cannot be explained by a deficiency in gluconeogenesis[124] since this pathway was found to have developed normally[123]. It could be explained by an inefficiency of glycogen release or by the smaller liver size and hepatic immaturity. The reduction in the amount of available glycogen can thus be accounted for, in spite of a normal concentration of glycogen per gram of liver parenchyma. On the other hand, the organ which consumes the most glucose, namely the brain, is proportionally larger in size[112, 116]. The postnatal hypoglycaemia could then be explained by this hepatocerebral disproportion. One would consequently expect that glycogen disappears more rapidly in the liver of the experimental animals than in the controls. Although one would suppose that, in order to remedy this hypoglycaemia, a maximum degradation of glycogen would occur, the rate of breakdown of glycogen has been found to be the same in both groups[15].

Another way of accounting for the hypoglycaemia could be the inability of the brain, as well as other organs of intrauterine growth retarded rats, to utilize ketone bodies[125]. This would force the various organs to use up more glucose, which results in a more rapid depletion of the glycogen reserves. In this regard, blood plasma concentrations of beta hydroxybutyric acid have been studied in both normal newborns and those affected by intrauterine growth retardation[126]. Its concentration was found to be the same in both groups. The concentration of the enzyme beta hydroxybutyrate dehydrogenase in the brain was also measured in the two groups and found to be equal[126].

These various hypotheses cannot entirely explain the hypoglycaemia which occurs at birth. It is possible that changes in the fetomaternal gradient are also responsible for this drop in glycaemia.

In conclusion, a relative incapacity for extrauterine adaptation has been induced by the factors causing intrauterine growth retardation. This incapacity can, therefore, be considered to be teratogenic: the newborn is abnormal as regards this particular function. Moreover, these anomalies could be definitive. For instance, the weight of these animals when in the adult state remains below normal[127].

The effect of prematurity and postmaturity on the development of enzymes is equally of interest. As has already been shown, certain functions commence towards the end of gestation.

Prematurity

The question can, therefore, be raised as to what is the impact on subsequent development when birth occurs before the onset of certain metabolic functions (for example, before phosphorylase or glucose-6-phosphatase becomes operational). In the case of prematurity, blood glucose concentrations are found to be much lower at birth. However, the evolution is completely similar to that of the normal neonate, but continues to be marked by the lag which started at birth[4,128]. The glycogen reserves are similar to those found in the liver of term births on the twenty-first day.

Breakdown of glycogen is possible, but it starts with a certain delay which is proportional to the degree of prematurity. Plasma concentration of lactate at birth is the same as in term deliveries. It is, however, only after 120 minutes that the concentration of lactic acid reaches a value close to that found in the adult[128].

As far as the enzyme modifications are concerned, it was shown that the induction of phosphoenolpyruvate carboxykinase occurred at birth in the same way on the twentieth day as after a normal gestation period[58,59]. However, both phosphopyruvate carboxylase and tyrosine amino transferase were induced and developed more slowly than in the animal born at term[4].

Although no major anomalies have been observed in the adaptation to extrauterine life it must be mentioned that only 24 h prematurity has been investigated. It is possible that greater prematurity could prevent normal adaptation.

Postmaturity

In postmaturity of three days induced by the injection of progesterone, one finds higher blood glucose values in the newborn[129,130]. However, the trend towards hypoglycaemia is the same as that observed in normal newborns[129]. The glycogen reserve is reduced as compared to that of neonates born at term[131]. Its mobilization occurs as in term animals. Glucose-6-phosphatase and phosphorylase differentiate normally after birth[132]. Lactacidaemia, on the other hand, is markedly higher and normal values are not observed within normal delays[129]. As far as gluconeogenesis and the induction of phosphoenolpyruvate carboxykinase are concerned it has been shown[133] that, although this enzyme may be induced *in utero* by prolonged gestation, the postnatal capacity for gluconeogenesis is not altered[16,132]. It has been suggested that the limiting factor of gluconeogenesis is the ratio of NAD/NADH which must be modified by the oxygenation that can only occur at birth[133]. The impact of a prolonged gestation on two enzymes of the urea cycle have also been studied[134]. It was shown that OCT develops spontaneously *in utero*, whereas arginase is induced by the birth process itself[134].

These observations indicate that in the adaptative metabolic development there are, on the one hand, changes which are induced by the birth process (the triggering factors could be the stress of delivery or of the postnatal environment such as temperature, anoxia and tactile stimuli) and, on the other hand, there are the processes of development which are unrelated to birth and which unfold according to a predetermined genetic programme.

Conclusion

Intrauterine growth retardation, by modifying fetal and neonatal metabolism, alters the development of the adaptative process. Consequently, it must be considered as a teratogenic event.

Prematurity constrains the newborn to initiate new metabolic functions with an immature enzyme equipment. Indeed, the birth process itself induces certain functions at a stage when these would be normally still inactive. Therefore, abnormal patterns of development are present and this may be considered as a teratogenic event.

In postmaturity some functions mature *in utero*. The metabolic function thus activated behaves *in utero* as it would during the postnatal period. This modifies the neonate's adaptative capacity. Here, too, the developmental process is modified and postmaturity can therefore be considered as a teratogenic event.

CHEMICAL AND HORMONAL FACTORS MODIFYING THE ADAPTATION TO EXTRAUTERINE LIFE

Preliminary remarks

A clear distinction must be made between factors that are purely toxic in nature and those that interfere with the developmental process itself. A factor that causes the necrosis of a fetal parenchyma, or inhibits a function that has already been established, cannot be considered teratogenic; it is merely toxic. An example of such a factor is the drop in prothrombin time observed after the administration of salicylate[135]. If, however, this factor interferes with a developmental process, it must be considered as teratogenic; for instance, the momentary inhibition of mitoses in the germinative layer of the embryonic brain will later result in the absence of certain well-defined cells[136] in the adult brain.

First, the effects of the excess or the absence of certain hormones on the process of differentiation will be studied. Then, some examples of chemical products that alter the development of certain enzymes intervening in the adaptation to extrauterine life will be examined.

The influence of corticoids on differentiation

Corticoids play an important role in the normal differentiation of various factors. These results were discovered by the study of the consequences of an excess or lack of hormone during gestation. Physiologists have thus been dependent on teratological experiments to find the exact role of certain hormones. Their results will be used to examine developmental problems but, before considering the effect of hormones administered to the mother on fetal development, it must be proven that these hormones cross the placenta. This was done for corticoids at least as far as late gestation is concerned[137]. It has been shown that they suppress the fetal adrenocortical system by their action on the hypothalamus–pituitary–adrenocortical axis[138-140]. After the administration of corticoids to the fetus, blood glucose is high and reaches normal values within the first few hours[141]. When administered from the seventeenth gestational day, the accumulation of liver glycogen between the eighteenth and twenty-first days is accelerated[12]. However, the final concentration of glycogen is similar to that found during normal differentiation. The administration of corticoids to the fetus from the sixteenth day, i.e. before the phase of rapid glycogen synthesis, results in its premature onset[142]. The effect on time of appearance of enzymatic activity has not been studied. As to the urea cycle, the effect of triamcinolone on arginine synthethase activity has been demonstrated[142]. This induction effect seems, however, to be rather subtle. Indeed, before birth no effect could be obtained *in vivo* except for carbamyl phosphate synthetase[144-146]. After birth a substantial effect on activity of arginine synthetase has been observed[143]. *In vitro*, corticoids are active on differentiation of the enzymes of the cycle (except OCT) as well in livers taken before as after birth[147].

Hormonal deprivation, by maternal adrenalectomy alone or by fetal adrenalectomy alone, was insufficient to demonstrate the role of fetal adrenals[18, 148]. However, maternal adrenalectomy associated with decapitation of the fetus resulted in the complete suppression of the fetal adrenal activity[149]. This technique was able to demonstrate the important metabolic modifications due to hormonal deprivation. The stores of glycogen in the liver were low[150]. It is interesting to note that, despite this absence of liver glycogen, fetal blood glucose did not seem to be modified[150]. However, these animals could not be studied during the postnatal period. The adrenalectomy performed immediately after birth was responsible for a decrease in the enzymatic activity of arginine synthetase[143].

A more thorough study of all five enzymes of the urea cycle was carried out[144] and showed that an antenatal fetal adrenalectomy resulted in a reduction of carbamyl phosphate synthetase, OCT and argininosuccinate synthetase. The administration of metopirone gave similar results to those of the adrenalectomy[151]. Corticoid deprivation causes a delay in the development of arginase, but is without effect on that of OCT[151].

Although extensive studies have been carried out on the consequence of hormonal deprivation on certain enzymes and on enzyme development, there is as yet little data concerning their effects on the metabolic aspects of the neonatal period, i.e. on the adaptation to extrauterine life.

The influence of thyroid hormones

It is generally admitted, although not yet established, that a transplacental passage of thyroid hormones is possible in the rat at the end of gestation[92]. It is, therefore, logical to start by examining the effects on the mother of the administration of these hormones. The administration of T_3 to the mother 24 h before sacrifice on the twentieth day of gestation caused a fall in the liver glycogen concentration[152]. The same result can be obtained by injecting T_3 into the fetus[153]. This fall could be attributed to the premature induction of glucose-6-phosphatase. These latter results should be interpreted with caution, however, since an anaesthesia is needed for intrauterine injections and this might in itself have an effect on the carbohydrate metabolism of the fetus[39]. Little is yet known concerning the effect of thyroid hormones on the enzymes of the urea cycle. This, one expects, will soon be established, since it has already been shown that in the tadpole the appearance of these enzymes is linked to the process of metamorphosis and may be delayed by a thyroid insufficiency[154]. A thyroid hormone analogue, 3,5 dimethyl-3'-isopropyl-L-thyronine (DIMIT) is capable of inducing certain hepatic enzymes, among others tri-iodothyronine amino transferase[155]. The oral administration of propylthiouracil after the ninth day of gestation inhibits its formation[155].

DIMIT has also been found to cause a great reduction in the hepatic glycogen content with a decrease of the enzyme glucose-6-phosphatase[156]. An increase in the number of mitochondria present in the hepatic cells was also noted. This could have serious repercussions on the adaptation to extrauterine life, since it has been postulated that the function of the Krebs cycle increases greatly at the moment of birth[24]. Mitochondria play an essential role in this cycle. Stimulation of the mitochondrial activity may lead to an increase in the oxygen needs with a subsequent decrease of the natural resistance to hypoxia during this period of relative anoxia. Maternal thyroidectomy before gestation causes a reduction of the glycogen reserves[152]; the other parameters of the neonatal period have not been studied[157]. Moreover, it is known that the fetal thyroid is not changed morphologically by the thyroidectomy of the mother, but its function appears to be modified and may affect later development[158-160]. It must be remembered that the thyroid gland greatly influences the functional development of the central nervous system (CNS)[161]. However, the CNS plays only a relatively unimportant role in the adaptation to extrauterine life, therefore the thyroid-CNS interrelation will not be further discussed here. There is, nonetheless, one particular aspect of this differentiation that can have metabolic repercussions in the neonatal period. Thyroid hormones affect the development of the pituitary gland and of plasma corticosterone diurnal rhythm[162-164]. This in turn may modify the adaptation to extrauterine life, since we have described the role of corticoids in this process.

The action of glucagon on differentiation

It has been suggested that glucagon may induce premature development of glucose-6-phosphatase[12]. There is an increase in the enzymatic activity of arginine synthetase after administration of glucagon[50]. However, this can only be obtained around the fifth postnatal day.

The effect of certain chemicals on perinatal development
Alcohol
Alcohol has attracted much interest due to its widespread use. Recently a 'fetal alcohol syndrome' has been described[165] but the biochemical and metabolic genesis of this syndrome are not well understood. A few metabolic studies have been attempted. Among these the synthesis of proteins in the fetal liver was investigated[166]. An important reduction was noted after the repeated administration of alcohol during gestation. The enzyme activity of tyrosine transaminase and of tryptophan oxygenase were not changed[160]. A study of the effect of alcohol on glycogen and blood glucose levels has also been made[167]. It has been shown that maternal alcohol consumption is responsible for alterations of ribosomal protein synthesis in the fetal and newborn brain[168].

Luminal
A single dose of Luminal, administered 60 to 80 minutes before Caesarean section on the twentieth or twenty-first day of gestation, decreases the amount of liver glycogen[39, 169]. This is most likely due to an activation of phosphorylase[39].

If Luminal is given between the tenth and fifteenth gestational day, it causes a reduction in the liver glycogen reserves between the sixteenth and the twentieth day[170]. Luminal given between the seventeenth and the nineteenth day gives the following results[169]. At birth glycogen content of the liver is normal; however, mobilization of this glycogen is impaired. Nevertheless hypoglycaemia is not observed. Therefore, it may be suggested that gluconeogenesis starts earlier in these animals than in controls. It is also noteworthy that brown fat glycogen increases but that its mobilization is also impaired.

Phenobarbital is known to be an enzyme inducer but, as of yet, no studies have been made on its effect on the induction of enzymes intervening in fetal liver carbohydrate metabolism.

Diphenylhydantoin
Diphenylhydantoin is another known enzyme inducer. Administered to the mother from the seventeenth to the nineteenth gestational day, it produces a nearly complete exhaustion of the glycogen reserves in the fetal liver. It is accompanied by major hyperglycaemia[171].

Salicylates
Salicylates administered to the mother modify the glycogen content of newborn mice[172]. This action, however, is not observed when salicylates are administered to rats. Indeed, the glycogen content remains unchanged[173]. Nonetheless, a slight hyperglycaemia is noted[173]. The amount of glycogen in brown fat is greatly reduced and thus may affect thermoregulation[173].

CONCLUSION

The neonatal period is a time of important morphological and functional changes. Considerable changes must occur in order to allow the newborn to assume new metabolic requirements. The neonatal period is, therefore, a challenge.

The adaptation to extrauterine life is rendered possible by the antenatal acquisition of certain functional potentialities, which will however only be operative after birth, and the establishment of new metabolic functions after birth.

The moment of birth constitutes a decisive turning point. Indeed, this event stimulates the *activation* of certain potentialities acquired before birth and the *induction* of new functions.

A disturbance in one of the developmental processes intervening in this adaptation will be responsible for the maladjustment to extrauterine life. It can, therefore, be considered as a teratogenic event. This inadaptation to extrauterine life can arise in the following circumstances: in the absence of the *antenatal induction* of an enzyme intervening in some postnatal metabolic process; in the absence of the *postnatal activation* of an enzyme already present during fetal life but in its inactive form; and in the absence of the *postnatal induction* of a new metabolic function.

The premature, antenatal differentiation of a function which normally only appears after birth can also be responsible for inadaptation. Prematurity and postmaturity, by displacing the moment of birth with respect to the normal developmental processes, result in the premature or postponed induction of certain metabolic functions. Disturbances in the adaptation process can lead secondarily to other developmental anomalies, which sometimes reveal themselves only much later. Some examples will illustrate how plastic the neonatal period is. Certain substances, when given during the particular critical period of organization, can alter the normal developmental process in later life. Sex hormones or sex hormone antagonists, when given during the perinatal period, can be responsible for the permanent impairment of the hypothalamo-pituitary axis. This will, however, only be manifest at the time of puberty[174,175]. Treatment with gonadotropic hormones during the neonatal period was found to be responsible for a decrease of the thyroid's thyroxine response to TSH[176]. Perinatal administration of thyroxine causes durable impairment of the hypothalamo-pituitary-thyroid axis[177]. Neonatal treatment with angiotensin is responsible for hypertension in adulthood[178]. The administration of L-dopa and of 5-hydroxytryptophan to neonatal rats causes persistent alterations in pituitary function[179]. Rats treated during the neonatal period with glucosamine showed an invariable decrease in the fasting blood glucose values and a permanent increase in the insulin values[180]. The glucose receptors on the beta cells appeared markedly reduced.

All these experiments emphasize a new aspect of teratology. In the same way as the morphology of organs can be changed, cell membrane morphology can also be modified. It has been shown that the cell membrane receptors can be altered by different drugs given during a certain critical period of development. Indeed, during the perinatal period the plasticity of these receptors is considerable and any modification of imprint may have permanent consequences. It has been shown that even slight metabolic modifications occurring during the neonatal period can have durable effects. These deleterious effects are sometimes only manifest later in life and one can speak of a deferred teratological action. They can also be secondary to the metabolic changes which result from altered adaptation to extrauterine life. It is, therefore, evident that the field of teratology must be extended to include the neonatal period and the functional modifications which can occur during the perinatal time.

References

1. Girard, J. R., Cuendet, G. S., Marliss, E. B., Kervran, A., Rieutort, M. and Assan, R. (1973). Fuels, hormones and liver metabolism at term and during the early postnatal period in the rat. *J. Clin. Invest.*, 52, 3190
2. Watts, C., Gain, K. and Sandin, P. L. (1976). Glucose homeostasis in the developing rat. I. Blood glucose and immunoreactive insulin in the later stages of gestation of the fetal rat. *Biol. Neonate*, 30, 88
3. Goodner, C. J., Conway, M. J. and Werrbach, J. H. (1969). Relation between plasma glucose levels of mother and fetus during maternal hyperglycemia, hypoglycemia and fasting in the rat. *Pediatr. Res.*, 3, 121
4. Di Marco, P. N., Chisalberti, A. V., Pearce, P. H. and Oliver, I. T. (1976). Postnatal changes in blood glucose, phosphopyruvate carboxylase and tyrosine aminotransferase after normal birth and premature delivery in the rat. *Biol. Neonate*, 30, 205
5. De Meyer, R., Gerard, P. and Verellen, G. (1971). Carbohydrate metabolism in the newborn rat. In: Jonxis, J. H. P., Visser, H. K. A. and Troelstra, J. A. (eds.) *Nutricia Symposium on Metabolic Processes in the Fetus and Newborn Infant*. p. 281.
6. Snell, K. and Walker, D. G. (1973). Glucose metabolism in the newborn rat. Temporal studies in vivo. *Biochem. J.*, 132, 739
7. Dawkins, M. J. (1963). Glycogen synthesis and breakdown in fetal and newborn rat liver. *Ann. N.Y. Acad. Sci.*, 111, 203
8. Yeung, D. and Oliver, I. T. (1968). Induction of phosphopyruvate carboxylase in neonatal rat liver by adenosine 3′,5′-cyclic monophosphate. *Biochemistry*, 7, 3231
9. Cake, M. H., Yeung, D. and Oliver, I. T. (1971). The control of postnatal hypoglycemia. Suggestions based on experimental observations in neonatal rats. *Biol. Neonate*, 18, 183
10. Gain, K. and Watts, C. (1976). Glucose homeostasis in the developing rat. II. Blood glucose, plasma insulin and hepatic glycogen in the newborn rat. *Biol. Neonate*, 30, 35
11. Blasquez, E., Sugase, T., Blasquez, M. and Foa, P. P. (1974). Neonatal changes in the concentration of rat liver cyclic AMP and of serum glucose, free fatty acids, insulin, pancreatic and total glucagon in man and in the rat. *J. Lab. Clin. Med.*, 83, 957
12. Greengard, O. and Dewey, H. K. (1970). The premature deposition or lysis of glycogen in livers of fetal rats injected with hydrocortisone or glucagon. *Dev. Biol.*, 21, 452
13. Girard, J., Ferre, P. and Gilbert, M. (1975). Le métabolisme energétique pendant la période périnatale. *Diabète Metabol. (Paris)*, 1, 241
14. Girard, J. R. and Zeghal, N. (1975). Adrenal catecholamines. Content in fetal and newborn rats. Effects of birth by caesarian section, cold exposure, hypoxia, hypoglycemia and 2-deoxyglucose. *Biol. Neonate*, 26, 205
15. Chanez, C., Tordet-Cardiroit, C. and Roux, J. M. (1971). Studies on experimental hypotrophy in the rat. II. Development of some liver enzymes of gluconeogenesis. *Biol. Neonate*, 18, 58
16. Ballard, F. J. (1971). The development of gluconeogenesis in rat liver. Controlling factors in the newborn. *Biochem. J.*, 124, 265
17. De Meyer, R., Verellen, G. and Niethals, A. M. (1979). Citrate concentration in blood of the newborn rat. (In preparation)
18. Jacquot, R. (1959). Research on the endocrine control of glycogen accumulation in the

liver of rat fetus. I. Experiences of decapitation of the fetus in utero. *J. Physiol. (Paris),* 51, 655
19. Ballard, F. J. and Oliver, I. T. (1963). Glycogen metabolism in embryonic chick and neonatal rat liver. *Biochem. Biophys. Acta,* 71, 578
20. De Meyer, R. (1979). The role of carbohydrates in energy balance of premature and term born rats. *Pediatr. Res.* (In press)
21. Burch, H. B., Lowry, O. N., Kuhlman, A. M., Skerjance, J., Diamant, E. J., Lowry, S. R. and Von Dippe, P. (1963). Changes in patterns of enzymes of carbohydrate metabolism in the developing rat liver. *J. Biol. Chem.,* 238, 2267
22. Stave, U. (1964). Age-dependent changes of metabolism. I. Studies of enzyme patterns of rabbit organs. *Biol. Neonate,* 6, 128
23. Hommes, F. A. and Wilmink, C. W. (1968). Developmental changes of glycolytic enzymes in rat brain, liver and skeletal muscle. *Biol. Neonate,* 13, 181
24. Hommes, F. A. (1971). Development of enzyme systems in glycolysis. In: Jonxis, J. H. P., Visser, H. K. A. and Troelstra, J. A. (eds.) *Nutricia Symposium on Metabolic Processes in the Foetus and Newborn Infant.* p. 3.
25. Greengard, O. (1977). Enzymic differentiation of human liver: comparison with the rat model. *Pediatr. Res.,* 11, 669
26. Vergonet, G. (1971). Computer representation of glycolysis in adult and foetal rat liver. In: Jonxis, J. H. P., Visser, H. K. A. and Troelstra, J. A. (eds.) *Nutricia Symposium on Metabolic Processes in the Foetus and Newborn Infant.* p. 14
27. Ureta, T., Bravo, R. and Babul, J. (1975). Rat liver hexokinase during development. *Enzyme,* 20, 334
28. Burch, H. B. (1963). Biochemical changes in the perinatal rat liver. *Ann. N.Y. Acad. Sci.,* 111, 176
29. Jamdar, S. C. and Greengard, O. (1970). Premature formation of glucokinase in developing rat liver. *J. Biol. Chem.,* 245, 2779
30. Walker, D. G., Khan, H. H. and Eaton, S. W. (1966). Enzymes catalysing the phosphorylation of hexoses in neonatal animals. *Biol. Neonate,* 9, 224
31. Walker, D. G. and Holland, G. (1965). The development of hepatic glucokinase in the neonatal rat. *Biochem. J.,* 97, 845
32. Vernon, R. G. and Walker, D. G. (1968). Changes in activity of some enzymes involved in glucose utilization and formation in developing rat liver. *Biochem. J.,* 106, 321
33. Hanson, R. W., Garber, A. J., Reshef, L. and Ballard, R. J. (1973). Phosphoenolpyruvate carboxykinase. II. Hormonal controls. *Am. J. Clin. Nutr.,* 26, 55
34. Ballard, F. J. and Oliver, I. T. (1962). Appearance of fructose-1,6-diphosphatase in postnatal rat liver. *Nature (Lond.),* 195, 498
35. Hommes, F. A., Luit-De Haan, G. and Richters, A. R. (1971). The development of some Krebs cycle enzymes in rat liver mitochondria. *Biol. Neonate,* 17, 15
36. De Vos, M. A., Wilmink, C. W. and Hommes, F. A. (1968). Development of some mitochondrial oxidase systems of rat liver. *Biol. Neonate,* 13, 83
37. Mäenpää, P. H. and Räihä, N. C. R. (1978). Content of cytochrome $(a + a_3)$ in the heart, liver and brain of the developing rat. *Scand. J. Clin. Lab. Invest.,* 21 (Suppl. 101), 7
38. Hommes, F. A. and Richters, A. R. (1969). Mechanism of oxidation of cytoplasmic nicotinamide adenine dinucleotides in the developing rat liver. *Biol. Neonate,* 14, 359
39. Devos, P. and Hers, H. G. (1974). Glycogen metabolism in the liver of the foetal rat. *Biochem. J.,* 140, 331
40. Weber, G. and Cantero, A. (1955). Glucose-6-phosphatase activity in regenerating, embryonic and newborn rat liver. *Cancer Res.,* 15, 679
41. Isselbacher, K. J. (1957). Evidence for an accessory pathway of galactose metabolism in mammalian liver. *Science,* 126, 652
42. Jacquot, R. and Kretchmer, N. (1964). Effect of fetal decapitation on enzymes of glycogen metabolism. *J. Biol. Chem.,* 239, 1301
43. Schaub, J., Gutman, I. and Lippert, H. (1972). Developmental changes of glycolytic and gluconeogenic enzymes in fetal and neonate rat liver. *Horm. Metab. Res.,* 4, 110
44. Okuno, G., Price, S., Grillo, T. A. *et al.* (1964). Development of phosphorylase and phosphorylase-activating (glucagon-like) substances in the rat embryo. *Gen. Comp. Endocrinol.,* 4, 446

liver. In: F. A. Hommes and C. J. Van Den Berg (eds.). *Inborn Errors of Metabolism*. (New York and London: Academic Press)
144. Gautier, C., Husson, A. et Vaillant, R. (1975). Evolution et contrôle de l'activité de la carbamyl phosphate synthetase et de l'ornithine transcarbamylase du foie foetal de rat au cours du développement. *Experientia*, 31, 125
145. Gautier, C., Husson, A. et Vaillant, R. (1977). Effets des glucocorticosteroides sur l'activité des enzymes du cycle de l'urée dans le foie foetal de rat. *Biochimie*, 59, 91
146. Husson, A., Gautier, C. et Vaillant, R. (1975). Evolution et contrôle de l'activité de l'argininosuccinate synthetase, de l'argininosuccinase et de l'arginase du foie foetal de rat. *Experientia*, 31, 1403
147. Räihä, N. C. R. and Edkins, E. (1976). Effects of dexamethasone, cyclic AMP and insulin on the activity of urea cycle enzymes in fetal liver in organ culture. *Pediatr. Res.*, 10, 884
148. Adam, P. A. J. (1971). Control of glucose metabolism in the fetus and newborn infant. In: R. Levine and R. Luft. (eds.) *Advances in Metabolic Disorders*. 5, p. 183 (New York and London: Academic Press)
149. Jost, A. (1966). Problems of fetal endocrinology: the adrenal glands. *Recent Progr. Horm. Res.*, 22, 541
150. Jacquot, R. (1959). Research on the endocrine control of glycogen accumulation in the liver of rat fetus. II. Experiences of decapitation of the fetus in utero. *J. Physiol. (Paris)*, 51, 693
151. De Meyer, R. (1978). Arginine and OCT in the liver of newborn rats after administration of metopirone to the mother.
152. Gommers, A. (1967). Recherches sur le rôle de la thyrolde maternelle dans le développement prénatal du rat. (Bruxelles: Ed. Arscia)
153. Greengard, O. and Dewey, H. K. (1968). The developmental formation of liver glucose-6-phosphatase and reduced nicotinamide adenine dinncleotide phosphate dehydrogenase in fetal rats treated with thyroxine. *J. Biol. Chem.*, 243, 2745
154. Brown, G. W., Brown, W. R. and Cohen, P. P. (1959). Levels of urea cycle enzymes in metamorphosing *Rana catesbeiana* tadpoles. *J. Biol. Chem.*, 234, 1775
155. Benson, M. C., Liu, J. P., Huans, Y. P., Burger, A. and Riulin, R. S. (1978). Differential effects of triiodothyronine and 3,5-dimethyl 3'-isopropyl-L-thyronine treatment of maternal rats on hepatic-L-triiodothyronine amino transferase activity in fetal rats. *Endocrinology*, 102, 562
156. Kriz, B. M., Jones, A. L. and Jorgensen, E. C. (1978). Effects of a thyroid hormone analog on fetal rat hepatocyte ultrastructure and microsomal function. *Endocrinology*, 102, 712
157. Battarbee, H. D. (1974). The effects of thyroid state on rat liver glucose-6-phosphatase activity and glycogen content. *Proc. Soc. Exp. Biol. Med.*, 147, 337
158. Stempak, J. G. (1962). Maternal hypothyroidism and its effects on fetal development. *Endocrinology*, 70, 443
159. Varma, S. K., Murray, R. and Stanbury, J. B. (1978). Effect of maternal hypothyroidism and triiodothyronine on the fetus and newborn in rats. *Endocrinology*, 102, 24
160. Bakke, J. L., Lawrence, N. L., Robinson, S. and Bennett, J. (1975). Endocrine studies of the untreated progeny of thyroidectomized rats. *Pediatr. Res.*, 9, 742
161. Nathanielsz, P. W. (1975). Thyroid function in the fetus and newborn mammals. *Br. Med. Bull.*, 31, 51
162. Lengvari, I., Branch, B. J. and Taylor, A. N. (1977). The effect of perinatal thyroxine treatment on the development of the plasma corticosterone diurnal rhythm. *Neuroendocrinology*, 24, 65
163. Lengvari, I., Branch, B. J. and Taylor, A. N. (1977). Effect of perinatal thyroxine treatment on some endocrine functions of male and female rats. *Neuroendocrinology*, 24, 129
164. Taylor, A. N. and Lengvari, I. (1977). Effect of combined perinatal thyroxine and corticosterone treatment on the development of the diurnal pituitary-adrenal rhythm. *Neuroendocrinology*, 24, 74
165. Jones, K. L., Smith, D. W., Ulleland, C. N. and Streissguth, A. P. (1973). Pattern of malformation in offspring of chronic alcoholic mothers. *Lancet*, i, 1267

166. Rawat, A. (1976). Effect of maternal ethanol consumption on foetal and neonatal rat hepatic protein synthesis. *Biochem. J.,* **160,** 653
167. De Meyer, R. (1979). Effect of maternal administration of alcohol on carbohydrate metabolism of the newborn rat. (In preparation)
168. Rawat, A. K. (1975). Ribosomal proteins synthesis in the fetal and neonatal rat brain as influenced by maternal ethanol consumption. *Res. Commun. Chem. Pathol. Pharmacol.,* **12,** 723
169. De Meyer, R. (1979). Effect of phenobarbital on carbohydrate metabolism in the fetal rat liver. (In preparation)
170. Delphia, J. M. and Singh, S. (1972). Effect of maternally injected sodium pentobarbital during the embryonic period of gestation on liver glycogen levels in the rat fetus. *Res. Commun. Chem. Pathol. Pharmacol.,* **4,** 7
171. De Meyer, R. (1979). Effect of maternal administration of diphenylhydantoin on carbohydrate metabolism in fetal rat liver. (In preparation)
172. Larsson, K. S. (1971). Action of salicylate on prenatal development. In: Tuchmann-Duplessis | (ed.). | *Malformations | Congénitales des Mammifères.* pp. 171–186 (Paris: Masson)
173. De Meyer, R. (1979). Effect of maternal administration of salicylates on carbohydrate metabolism in fetal rat liver. (In preparation)
174. Dörner, G. (1974). Environment-dependent brain differentiation and fundamental processes of life. *Acta Biol. Med. Germ.,* **33,** 129
175. Shapiro, B. H., Goedman, A. S. and Root, A. W. (1974). Prenatal interference with the onset of puberty, vaginal cyclicity and subsequent pregnancy in the female rat. *Proc. Soc. Exp. Biol. Med.,* **145,** 334
176. Csaba, G. and Nagy, S. U. (1976). Plasticity of hormone receptors and possibility of their deformation in neonatal age. *Experienta,* **32,** 651
177. Lammers, M., Von zur Mühlen, A., Döhler, V. (1973). Prenatal thyroxine treatment causes permanent impairment of hypothalamo–pituitary–thyroid function in rats. *Acta Endocrinol.,* **87** (Suppl.), 215
178. Dörner, G., Götz, F., Hecht, K. and Hecht, T. (1976). Hypertension in adult rats treated neonatally with angiotensin. *Endokrinologie,* **68,** 231
179. Bakke, J. L., Lawrence, N. L., Robinson, S. A., Bennett, J. and Bowers, C. (1978). Late endocrine effects of L-dopa, 5-HTP and 6-OH-dopa administered to neonatal rats. *Neuroendocrinology,* **25,** 291
180. Csaba, G. and Dobozy, O. (1977). The sensitivity of sugar receptors. Analysis in adult animals of influences exerted at neonatal age. *Endokrinologie,* **69,** 227

13
Vital fluorochroming as a tool for embryonic cell death research

B. MENKES, OLTEA PRELIPCEANU AND I. CÀPALNÀŞAN

GENERAL INTRODUCTION

The zones of degeneration and cell death present in the embryo are well known from their first descriptions by Ernst[1] and Glücksman[2]. The existence of necrobiotic cells surrounded by intensely growing and even differentiating tissues was rather surprising and puzzling for many scientists. Yet, it soon became obvious that the appearance of these zones is a normal ontogenetic event with a precise topography with an onset and disappearance limited in time and space.

Glücksman[2] worked out a real 'inventory' of necrotic zones in various species. On the other hand, Menkes et al.[3,4] established a topographic relationship for several necrotic zones in human, rat, mouse, rabbit and avian embryos, proposing the term 'homologous zones'.

Glücksman[2] made an attempt to classify these zones, by distinguishing histogenetic, morphogenetic and phylogenetic degeneration. However, the actual mechanism and causal factors involved are so far unknown.

Several hypotheses have been put forward in order to explain the nature and the genesis of necrotic zones. 'Necrobiotic' metamorphosis in insects and in amphibia suggested a hormonal mechanism. Genetic 'lethal factors' have been also incriminated.

We tend to consider embryonic cell death in higher vertebrates a kind of recapitulation of necrobiotic reorganization during the metamorphosis of invertebrates and lower vertebrates. During this stage embryonic organs are reorganized or may even disappear.

Normal regression of some embryonic organs (Müllerian and Wolffian duct a.o.) involves typical necrobiosis and cell death.

A considerable amount of experimental work has been performed in order to elucidate the mechanism of the 'death clock' for a part of the embryonic cell population[5]. In spite of several approaches by heterochronic, homeo- or heterotypic graftings[5-8] by damaging the mesenchyme with cytotoxic substances, no essential progress has been made in our understanding of the genetic and epigenetic factors determining normal degenerative phenomena.

Embryonic cell death is frequently accompanied by a marked macrophage reaction. The macrophages, attracted to, or resulting from the transformation of neighbouring mesenchyme cells phagocytize the necrotic, most frequently nuclear debris. In our laboratory the theory of Chèvremont and Chèvremont-Comhaire[9] on the role of choline in the genesis of macrophages from mesenchyme cells has been verified[10]. Digestion is accomplished with the aid of lysosomal proteolytic enzymes. Indeed, a necrotic area is evident microscopically, first of all by these macrophages crowded by nuclear debris, and not by the primary necrotic particles.

In embryonic organs without or with little mesenchyme, the macrophage reaction is rather weak (e.g. in the central nervous system). It is obviously an argument for the mesenchymal local provenance of macrophages and against their accumulation via the bloodstream.

We observed a partial inhibition of the macrophage reaction with consecutive syndactyly induced by intoxication with Janus green[11,12]. This was observed in the interdigital zones of necrosis of the chick embryo and has recently been studied by several authors[8,13-16].

In our laboratory, the existence of about 60 normal, symmetrical necrotic zones in the human embryo has been established[17]. They have a precise topography and appear during certain periods ('phase specificity'). These necrotic areas (such as those detected in the facial processes, branchial arches, interdigital zones, retina, mesenchymal axial organs, heart, etc.) are frequently the starting point of some typical malformations.

Malformations in spontaneous and experimental teratogenesis are frequently initiated − morphologically − by the enlargement of one or more normal necrotic zones, or even by the appearance of new necrotic zones, absent in the normal embryo[3,11,18-20,59]. It can be stated that the influencing of the normal degenerative zones by exo- or endogenous agents may be teratogenic[3,11]. Thus, several mutants of the vertebral column are accompanied by abnormal necrotic processes. Spontaneous malformations or those induced by various factors: physical (hypoxia, irradiations), chemical, biological (virus infections[21]) may begin with the appearance of necrotic zones in the area of future malformations. Pexieder[22] reported in this respect some interesting observations. Following intra-amniotic administration of 30µg cyclophosphamide or 1 µg dexamethasone phosphate in 4-day chick embryos, malformations of the cardiac bulbus and of the atrioventricular areas are found, preceded by quantitative changes of the macrophages in the normal necrotic zones of the bulbus and of the areas mentioned. The author suggested an elective influence on these normal necrotic areas by the teratogens applied ('target effect').

Cytotoxic, cytostatic and antimitotic substances that are used in the treatment of malignant tumours induce in the embryo large necrotic zones showing cytological features identical to those found in normal necrosis. For example cyclophosphamide[23] induces in embryonic tissues growth retardation and large ubiquitous necrotic areas but does not prevent the macrophage reaction. In experimental alcoholic embryopathy[24], necroses precede the appearance of anomalies. Necrosis can be observed in certain immune reactions (e.g. in Rh or ABO incompatibility). We consider these reactions an expression of autoimmunity. Finally it can be assumed, that the reactive phenomena of the embryo may constitute a valuable model for dynamic investigations with respect to the effect of antitumoral agents upon human tumours.

It follows from the evidence outlined above, that the accumulation of knowledge concerning embryonic cell death, its repartition within the embryo, its role in normal morphogenesis and in spontaneous or experimental maldevelopment, and its use for screening the effects of antitumoral drugs has an outstanding theoretical and practical importance.*

The morphology of necrobiosis and necrosis in embryonic and fetal tissues has been extensively studied[5, 25]. In the necrotic areas cell death is always preceded by a relatively slow necrobiosis, lasting about 8 h. Changes seem to begin in the nucleus, which successively shows juxtamembranous hyperchromatosis, pycnosis and karyorhexis. Besides the nuclear changes, alterations of the cytoplasmic organelles are to be considered (frequently related to lysosomal activity). Sometimes nuclear chromatin is eliminated and 'worked up' intracytoplasmatically by lysosomal proteolytic enzymes (autophagy). The lysosome concept[26] is generally utilised for the understanding of the necrobiotic process.

It is sometimes difficult to ascertain which of the organelles are primarily affected. Several studies[15, 29, 30] have indicated mitochondrial changes. Electronmicroscopically[25], almost all the organelles of the 'sick' cells are affected. These cells present a positive acid phosphatase reaction[31, 32]. The cytoplasmic 'necrospherules' − very characteristic at the light-microscopic level − frequently contain mitochondrial fragments and are necrotic cytoplasmic areas. However, the intra- or extracytoplasmic (by phagocytosis) origin of the 'necrospherules' is sometimes difficult to establish.

Necrobiosis, if not still reversible, ends with necrosis. The cell debris thus formed contains both DNA and RNA (demonstrated by Feulgen and Brachet reaction).

METHODS OF DETECTION OF NECROTIC CELLS OR AREAS IN EMBRYONIC TISSUES

Postvital (on fixed preparations), supravital and even vital methods are applied. On serial sections stained by routine methods, intra- and extracellular 'necrospherules' are easily recognized. Pycnosis, karyorhexis, phagosomes are intensely basophilic. DNA and RNA content is detected by the method of Feulgen and Brachet. Fluorochroming of fixed preparations provided some results in the detection of mucopolysaccharides, but was rather inefficient for the detection of necrosis. Wendler[28] described in necrobiotic nuclei of fixed and fluorochromed preparations change in colour of the emerald green fluor-

---

* Among the contributions presenting exhaustive data as to embryonic cell death the following are quoted: cell death in normal vertebrate ontogeny[2]; death in embryonic systems[7]; cell death during embryogenesis[27]; cell death in teratogenesis[20]; embryonic and fetal cell death *in vivo* and *in vitro*[28]; cell death in the morphogenesis and teratogenesis of the heart[22]

escence (as it appears in normal nuclei) towards yellow and orange. The author described also the fluorescence of phagosomes on fixed and fluorochromed preparations. The results are not yet sufficient for the detection of necrotic zones. With the electronmicroscope intracytoplasmic necrotic elements are easily demonstrated as irregular, darkly contrasted masses. These features, corresponding either to phagocytized necrotic fragments (phagosomes) or autophagic necrotic zones (autophagosomes) of internal origin, are well known and need no further discussion.

On fixed and (preferably) unstained sections, phase-contrast microscopy reveals dense 'necrospherules'. The same method gives also excellent results with supravital, squashed preparations.

It is well known that particles (necrobiotic or necrotic remnants) show elective supravital staining. Such intra- or extracellular particles can achieve — from very diluted solutions (1 : 10 000, 1 : 50 000) — concentration of 1 : 100. We could verify this potency by cytospectrophotometry of Nile blue on macrophage phagosomes. No satisfactory explanation for this affinity exists as yet. The elective staining of nuclear remnants phagocytized by macrophages in degenerating zones of chick embryo limbs by Nile blue solutions of 1 : 10 000 or 1 : 20 000 have been introduced by Saunders et al,[6]. The 'Saunders-staining' gives excellent results for demonstrating superficially situated necrotic zones (subepidermal zones — interdigital, PNZ, facial zones, degenerating areas in dermatomes, branchial arches, etc.). The dye does not penetrate deeply and gives the best results on hypoxic or anoxic embryonic tissues. Whole embryos or detached limbs are immersed in the dye solution and controlled after about 30 minutes. This staining is not permanent and is difficult to fix.

We tried administering to the living chick embryo, through its blood circulation, neutral-red or Nile blue. A solution of 1 : 1000–1 : 2000 was applied onto the area vasculosa or onto the chorioallantoic membrane, in 5-, 6-, 7- and 8-day-old embryos. A diffuse staining of the embryonic body is obtained and some necrotic zones (in the heart at 5 days, interdigital at 7–8 days) are electively stained[33,34]. In these vital conditions however, staining is less marked than the supravital.

Significant progress has been made in our laboratory by the use of vital treatment with fluorochromes (acridine orange, choriphosine, quinacrine, auramine). Vital staining by fluorochromes has often been used, e.g. for investigating the concentration–potency of nephrons in amphibia, by injecting fluorescein into their lymphatic sac[35].

All necrotic zones may be located after 30 minutes, by fluorescence microscopy, introducing through the area vasculosa or chorioallantoic circulation one of these substances in dilutions of 1 : 5000, 1 : 10 000, 1 : 50 000 or even 1 : 100 000 (0.5 ml in physiological saline dropped onto the membrane). In the developing 4 day hen's egg, windowed as usual, 0.5 ml of 1 : 10 000 acridine orange or other flurochrome is dropped onto the area vasculosa. After 30 minutes the embryo is exteriorized and compressed between slide and coverslip or between two slides. This manipulation requires some experience. Dissociation of various organs or organ systems is obtained and folding of the tissues is avoided. Thus, accurate localization during examination of fluorescent zones are possible. On the other hand, appreciable thickness of the preparations reduces diffusion of excitable light which would interfere with the examination. In older embryos (after 7 days in chick embryo) dissection and separation of various organs and parts are recommended. In rodents (rats and mice between 12–16 days of pregnancy) the pregnant animal is anaesthetized with nembutal and laparotomized. After incision of the uterine swellings, embryos are partially exteriorized and injected intracardially using a Pasteur pipette. Continuation of heartbeats is one of the main conditions of successful injection. The heart may be warmed by dropping warm physiological saline during the intervention. During the injection the connection of the embryo with its placenta is preserved. Vascular diffusion of the fluorochrome is controlled by the gradually appearing yellowish colour of the embryo. About 30 min after the injection, the embryos are isolated, compressed (as in the case of chick embryos), and prepared for fluorescence microscopy. In mammals, the fluorochromes may be introduced by intravenous (into the tail vein) or by intraperitoneal injections.

Fluorescence is checked by a fluorescence microscope (UV light source HBO 200W, exciting filter blue 2XBG 12 Zeiss, special UV condenser and yellow protecting filter GG9 Zeiss). Photographs are made on black and white or colour film.

In total preparations, all intra- or extracellular necrotic remnants are intensely fluorescent, especially the phagosomes of the macrophages (Figures 13.1a, b)[33,34]. The organs containing mucopolysaccharides are also fluorescent (the fluorescence of the mucopolysaccharides after the treatment with acridine orange was reported by Saunders[36]). After intravital injection of flurochromes at the concentrations used in our experiments, on examination,

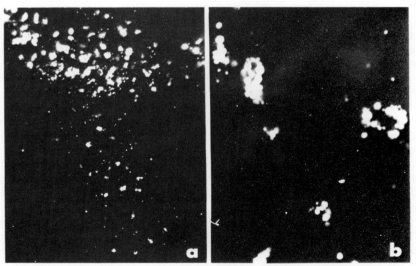

Figure 13.1 a, Chick embryo, 7.5 days of incubation. Vital fluorochroming by acridine orange. Interdigital space. Macrophages with intensely fluorescent phagosomes. Scattered, not yet phagocytized particles. b, Phagosomes of the macrophages

nuclei and lysosomes do not appear fluorescent at supravital examination, excepting the mesonephros which concentrates the injected dye. Its glomeruli show red–orange granules, the nuclei are greenish, and the intracytoplasmic lysosomes appear orange. An intense fluorescence of the nuclei and lysosomes is always obtained with high concentrations of fluorochromes, usually by direct application. For our purposes of detecting necrotic zones, however, this nuclear and lysosomal staining is rather inconvenient.

It is worth mentioning, that phagosomes stained supravitally by Nile blue cannot be secondarily stained by acridine orange because after fluorochroming, staining by Nile blue is not prevented.

Alternating examination of the same cell or cell group by both phase-contrast and fluorescence is also very useful (Figure 13.2). We wish to call

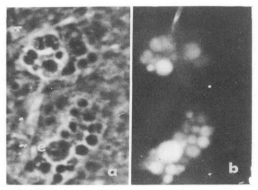

Figure 13.2 Comparison between macrophages examined by phase contrast and in ultraviolet light. a, Phagosomes demonstrated by phase-contrast. b, The same cells after treatment with acridine orange (note the unequal fluorescence of the phagosomes)

attention to the appearance in the 5 day chick embryo of a great number of migratory cells containing birefringent droplets. The relationship between these cells and macrophages are not yet established.

In the method described, fluorochromes are vitally administered and the examination is supravital. Comparing the various methods used for the detection of necrotic zones the following can be stated:

1. The 'Saunders staining' gives excellent results for the demonstration of superficial necrotic zones; it is technically simple and does not require special equipment.

2. Postvital study of fixed sections allows a precise localization of the necrotic cells and the application of histochemical methods. It offers, however, no possibility for obtaining a general survey of cell death throughout the embryonic organism.

3. Only vital application of fluorochromes via the bloodstream (and diffusion) provides – according to our experience – the possibility of determining the real amount of normal and induced necrotic processes in vertebrate embryos. The method requires special equipment and some experience in the manipulation of unfixed material.

When interpreting the findings it must be taken into account that mucopolysaccharides are also stained by fluorochromes, but without attaining the intensity of necrotic cells. On the other hand, some tissue and cells have their own fluorescence, thus a control on unstained preparations is always necessary.

4. An important advantage of the intravital administration of fluorochromes, which show relatively low toxicity and are well tolerated by the embryos at the concentrations used in our experiments, is the possibility of carrying out 'chronic' experiments. After a special treatment with fluorochromes, but appeared in the meantime, concentrate the excessive circunecrotic zones which did not exist at the time of application of the fluorochromes, but in the meantime now appeared, concentrate the excessive circulating fluorochrome or the substance persisting in the interstitial fluid and become fluorescent. The fixing of fluorochromes on necrotic elements continues several days after treatment. Administration may also be repeated on the same embryo.

During examination in ultraviolet light rapid photochemical changes occur, e.g. the 'extinction' of orange colour. This phenomenon must also be taken into account, especially when colour-film photography results for long exposure times are being considered.

THE EVIDENCE FOR SOME NORMAL NECROTIC ZONES IN THE CENTRAL NERVOUS SYSTEM

Hamburger and Levi-Montalcini[37] described in the chick embryo a constant necrotic zone in the cervical region of the spinal cord. According to Levi-Montalcini[38] the necrotized neuroblasts belong to the preganglionar visceral group of the ventrolateral column which disappears in the cervical area by a degenerative process between days 4.5 and 5 of incubation. Wechsler[39] found this necrosis on electronmicrographs of the anterior horn in chick embryos of 4.5 days. Jacobson[40] pointed out that the penetration of blood capillaries into the neural tube[41-43] coincided with the onset of this degeneration. The cellular debris, resulting from this necrosis would be – according to Levi-Montalcini and Hamburger – phagocytized by macrophages present in the bloodstream. Based upon our own observations we must confirm this opinion.

Necrosis in the spinal ganglia was also studied by Hamburger and Levi-Montalcini[37]. According to these authors, necrotic cells do not appear in the ganglia before 4.5 days of incubation. Cell death is maximal at 5–6 days, it decreases afterwards, and is completely absent at day 7. As stated by the above mentioned authors, no necrosis is detected in the ganglia which belong to the limbs. Necrotic cells are situated in the ventrolateral region of the ganglia where large early differentiating neuroblasts are found.

Menkes et al.[4] found necroses in the spinal ganglia of mouse, rat and human embryos. All the observations mentioned have been made on serial sections where routine staining clearly demonstrates the localization of necrobiosis and necrosis.

It is possible to make a general survey of necrosis in large segments of the spinal cord and along successive developmental stages with the vital fluorescence method (Figures 13.3 and 13.4). The neural tube is isolated under a

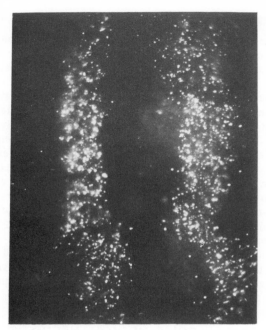

Figure 13.3 Chick embryo, 5 days of incubation. Vital fluorochroming by acridine orange. Cervical neural tube. Compression. Both motor columns show an intense fluorescence of necrotic remnants

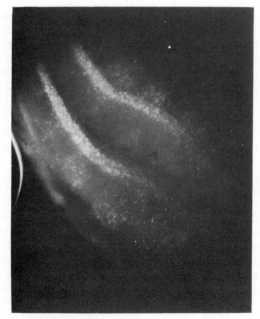

Figure 13.4 Chick embryo, 5 days of incubation. Vital fluorochroming by 1:5000 acridine orange. Compression. Necroses in the motor columns of the cervical spinal cord (the caudal end of the necrotic area is to be observed)

dissecting microscope and opened like a book along its dorsal suture. With the aid of a coverslip the two halves are slightly compressed.

Under the fluorescence microscope (low power examination) the two anterior motor columns appear as two luminary bands, composed by discrete, intensely fluorescent elements. Here, the extent of the necrotic process may be evaluated at the first glance. Necrosis involves the cervical zone and decreases stepwise towards the thoracic zone. In the anterolateral horns no necrosis is detected. At higher magnification, it is to be observed, that the intra- and extracellular fluorescent particles did not induce a uniform macrophage reaction as in other necrotic areas rich in mesenchyme. Isolated macrophages are to be found especially at the periphery of the necrotic areas, where cells of the perineural mesenchyme may have undergone transformation. As to the macrophages, our findings support their local origin from transformed mesenchymal cells. In the central nervous system, where mesenchyme is poorly represented their number is low.

In the day 7 chick embryo in the sacral spinal ganglia a clearcut ventrolateral necrosis is found (Figure 13.5). In some interesting cases (Figure 13.6),

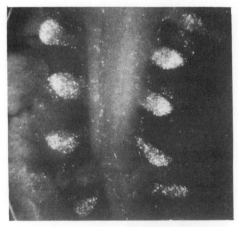

Figure 13.5 Chick embryo, 7 days of incubation. Vital fluorochroming by acridine orange. Lumbosacral spinal cord with spinal ganglia. Fluorescent necrotic particles in the ventral part of the ganglia

dorsal radicular fibres are distinguishable together with the ganglion and with the dorsal radix proper. Besides the above mentioned necrosis within the ganglion, necrotic elements are present along the dorsal root toward its junction with the ventral root and (a few but consistently) in the juxtamedullar portion of the dorsal radicular fibres.

Similar observations wer made in the day 15 rat fetus after injection of the acridine orange solution into the still beating heart. In Figure 13.7 intensely fluorescent necrotic elements are seen in the radicular fibres, close to the spinal cord. Necrosis is here localized in migratory cells, passing through the dorsal root toward the spinal nerve, the ramus communicans and the paravertebral ganglion chain. Are there cells migrating toward the spinal cord? A slight

ABNORMAL EMBRYOGENESIS: CELLULAR AND MOLECULAR ASPECTS

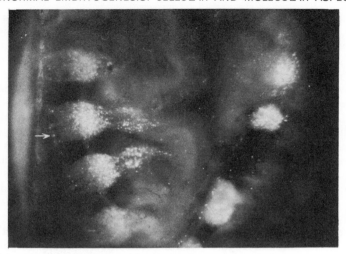

Figure 13.6 Chick embryo, 6 days of incubation. Vital fluorochroming by acridine orange. Strong compression. Thoracic spinal ganglia. Fluorescent necrotic particles in the dorsal radicular fibres (→)

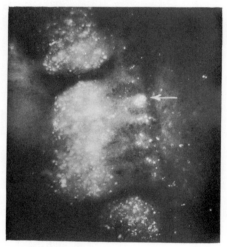

Figure 13.7 Rat embryo, 15 days of pregnancy. 1:10000 acridine orange intracardially. Spinal ganglion. Necrosis in the dorsal radicular fibres (→) and in the ventrolateral part of the ganglion

diffuse fluorescence within the ganglion may be due to some mucopolysaccharide content.

THE STUDY OF EMBRYONIC PREVERTEBRAL AXIAL ORGANS BY THE VITAL FLUORESCENCE METHOD

By using the previously described techniques in the chick embryo some intriguing features are seen, combining fluorescence of necrotic cells and of

VITAL FLUOROCHROMING IN CELL DEATH RESEARCH

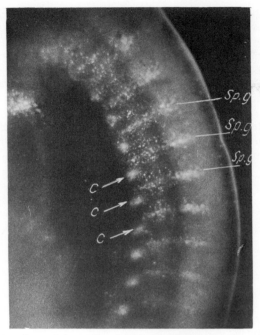

Figure 13.8 Chick embryo, 4 days of incubation. Vital fluorochroming with acridine orange (compression). Lateral view – thoracic area. Spinal ganglia with dorsal roots (Sp.g). Necrotic cells in the sclerotomes. Prevertebral 'centres' (C) alternating with fluorescent spinal nerves

mucopolysaccharides synthesized by sclerotome and chordal cells in various developmental phases. Thus, in the day 4 chick embryo a complex picture is obtained (Figure 13.8) in which metamerically ordered spinal ganglia and dorsal roots are evident. Alternating with the latter, ventrally from the chorda in the median plane, spherical fluorescent areas appear (prevertebral 'centres'). They are at this stage of development the first accumulation of mucopolysaccharides, a chemodifferentiation which precedes the morphogenesis of cartilaginous vertebral bodies. In the cervical zone the chordal membrane, the mucopolysaccharides of the future vertebral arches and, ventral from the chorda, necrotic sclerotomic cells are fluorescent.

In a recent paper Saunders[36] called attention to the possibility of visualizing the mucopolysaccharides in fixed, fluorochromized preparations. We also found (see above) fluorescence due to mucopolysaccharides but its intensity does not attain that of necrotic zones. Thus, the combined pictures can be easily interpreted.

The fluorescent centres described are stained by Alcian blue and precipitate the silver if impregnated (Aghdur method). They represent the areas of chondrogene sclerotomites and are checked on our photographs during or just after their reorganization. The supravital fluorescence method detects also very clearly the necroses in the dermatomes and in the ventral somitic crests.

The fate of dermatomes is especially interesting. Together with the loss of quasi-epithelial contacts of the lateral somite wall and the dispersion of the cells in the subepidermic space, an important rarefaction of this cell population occurs by selective necrosis and marked macrophage reaction. In the 4 day chick embryo this process involves initially, all the somites and persists longer in the more caudal areas, reaching a maximal intensity in the most caudal one. The necrosis in the dermatomes and the following macrophage reaction can be detected also by the Saunders staining (Figure 13.9a, b). In the 5 day chick embryo (Figure 13.10a, b) fluorescence of these zones is particularly marked (macrophages, loaded with phagosomes are very numerous). They present a metameric topography and preserve their connection with the corresponding somite levels. Showing a pyramidal form, with a dorsally oriented tip, their base is confluent with the ventral somitic crests

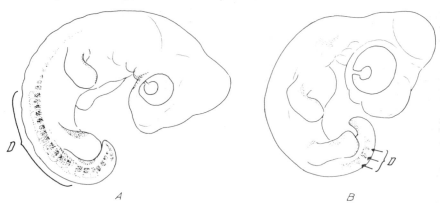

Figure 13.9 Chick embryo (A = 5 days; B = 6 days of incubation). Supravital staining with Nile blue 1 : 10 000. D = dermatomes

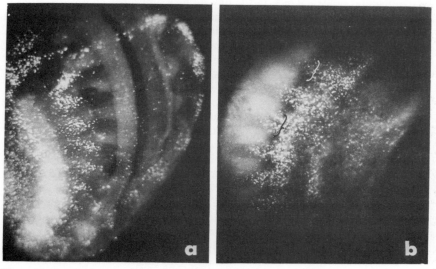

Figure 13.10 Chick embryo, 5 days of incubation. Vital fluorochroming by acridine orange. a, Caudal part of the body. b, Necroses in the dermatomes ({)

and with the neighbouring dermatomes.

On routinely stained, serial sections, the macrophages of the dermatomes are easily seen, but their global spatial arrangement becomes obvious by the fluorescence method. Our method demonstrates that the dermal mesenchyme retains, after its formation a segmental structure which depends on its somitic origin.

In the caudal zone, ventrally from the fluorescent dermatomes, a large mass of necrotic cells is evident. It corresponds to the urogenital region (Figure 13.11) in which extensive morphogenetic reorganizations involving degenerative processes are occurring. The tip of the caudal bud is also necrotic.

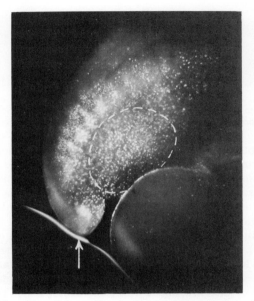

Figure 13.11 Chick embryo, 5 days of incubation. Vital fluorochroming by acridine orange. Necrosis of the caudal bud (→), of the urogenital zone (– – – –)

The possibility of visualizing the mesonephros (Figure 13.12) with its entire tubular system, including the Wolffian duct, should be pointed out. The fluorescence is due to a concentration of fluorochromes administered via the bloodstream. The glomeruli appear orange and the tubules emerald green. The concentrated fluorochrome demonstrates the nuclei and lysosomes.

Necrotic zones in the branchial arches, in the bulbus cordis and other areas are also demonstrated; examination by fluorescence completes the pictures obtained by the Saunders staining technique.

An important degenerative process in higher vertebrates, including man is localized around the chorda[44]. We analysed these perichordal necrosis in rat and mouse embryos (12–15 days of pregnancy respectively). The embryos were treated and examined as described above. The notochord of mouse and rat embryos, fixing the fluorochrome, shows a fluorescence of middle inten-

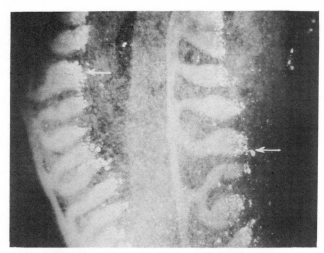

Figure 13.12 Rat embryo, 14 days of pregnancy. 1:10000 acridine orange intracardially. Strong compression. Mesonephros. Wolffian duct and nephrons (basal membranes and apical borders) show fluorescence. Intensely fluorescent granules at the glomerules (→)

sity. The sclerotomes are less fluorescent than the notochord, but the most impressive finding is a mantle of necrotic and necrobiotic cells around the thoraco-abdominal segment of the notochord, which also induces a localized macrophage reaction. These elements differ from the chordal and sclerotomal mucopolysaccharides by their fluorescence. The perichordal necrotic zone shows very obvious regular and metameric enlargements at the level of dense sclerotomites still existing at this stage of development (Figures 13.13a, b, c, d, e, and 13.14). Necrosis generally reaches to the chord membrane, but dispersed necroses are also found within the chord. In our laboratory, Sandor and Amels[45] observed enlarged perichordal necrotic areas in rat embryos at 14 days of pregnancy after treatment with 6-mercaptopurine. This phenomenon seems to be related to the biochemical predifferentiation which precedes chondrogenesis. A large number of studies indicate the morphogenetic role of the notochord in the formation of the vertebral column. The notochord precedes phylogenetically the appearance of vertebrae. In tissue culture, explanted somites show chondrogenesis, only when a fragment of the notochord is added to the culture. The excision of the notochord along several metameric segments prevents the segmentation of the vertebral column, the formation of separate vertebral bodies[46-54, 60].

The notochord thus participates as an inductor in the metameric organization of the skeletal axial organs, being present during the differentiation of each sclerotome towards a dense and a loose sclerotomite respectively. These sclerotomites undergo a reorganization and alternate with the myotomes, which eventually will be inserted onto two successive vertebrae. This complex process is not yet sufficiently clarified. A part of the dense segments takes part in the segmentation of the intervertebral discs; the loose segments contribute essentially to the shaping of the cartilaginous model of the vertebrae. The

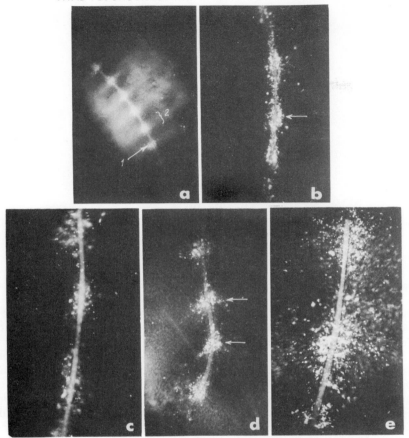

Figure 13.13 a, Rat embryo, 14 days of pregnancy. 1 : 10 000 acridine orange intracardially. Notochord with perichordal, segmentally enlarged necroses (1). Fluorescence of the dense sclerotomites (2). b, Rat embryo, 13 days of pregnancy. 1 : 10 000 acridine orange intracardially. Notochord with perichordal necrotic particles, segmentally accumulated (→). Intrachordal necrotic particles. c, Rat embryo, 13 days of pregnancy. 1 : 10 000 acridine orange intracardially. Notochord with perichordal, segmentally enlarged necroses. d, Rat embryo, 14 days of pregnancy. 1 : 10 000 acridine orange intracardially. Strong compression. Notochord and sclerotomic cells in various stages of necrobiosis and necrosis are fluorescent. Metameric disposition of perichordal necrosis (↑↑). e, Rat embryo, 13 days of pregnancy. 1 : 10.000 acridine orange intracardially. Notochord and sclerotomic cells show moderate, the perichordal necrotic particles intense fluorescence

chondroblasts secrete various mucopolysaccharides for the dense and for the lax sclerotomites. Menkes and Checiu[55] reported that the chondrogenic sclerotomites are Alcian blue positive before the appearance of cartilage, the dense sclerotomites being PAS-positive.

The metameric enlargements of the perichordal necrotic zones, described above, are in general corresponding to the PAS-positive dense sclerotomites. In our opinion, there exists a correlation between normal vertebral morphogenesis and the perichordal necrosis. It would be interesting to apply the

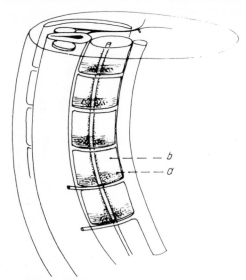

Figure 13.14 Scheme: The pattern of perichordal necrosis on the day 14 rat embryo (note the dense (a) and lax (b) sclerotomites)

method of global detection of perichordal necrosis in the mutants described in mice[56, 57], showing vertebral anomalies sometimes quite similar to the malformations of the human vertebral column.

In the foregoing, we emphasized the demonstration of normal necrotic areas by vital fluorescence. In our laboratory, this method has also been successfully used for the visualization of necrosis produced by chemical cytotoxic agents used in the treatment of malignant tumours. We present below the results obtained by fluorochroming embryos treated with cyclophosphamide, a cytotoxic substance of the 'nitrogen-lost' series that is used in cancer chemotherapy. The cytotoxic effects of this substance have been studied in detail by von Kreybig[23] who described its necrotizing effect in the rat embryo, following intraperitoneal or intravenous administration to pregnant animals. Similar findings were reported by Wendler[28]. As shown by these two authors, necrosis induced by cyclophosphamide is predominantly localized in the central nervous system and in the limbs and is followed by malformations.

When 100 μg of the substance (in physiological saline) was dropped upon the area vasculosa of chick embryo at 4 days incubation, after 24 h all the embryos were still alive, at 48 h mortality was almost 100%. When the dose was lowered to 50 and 10 μg, respectively, no death occurred.

Stained by supravital Nile blue (Saunders) 24 h after the administration of 100 μg cyclophosphamide the embryos showed intensely stained macrophages in the whole subepidermal mesenchyme (Figure 13.15), including the limbs, where the macrophages cover the normal marginal necrotic zones in the 5 day chick embryo.*

* Webster and Gross[58] reported the appearance of 'stippling' in embryos treated with mitomycin C (or by other agents inducing covalent cross-links in DNA) and stained afterwards by Nile blue or neutral red

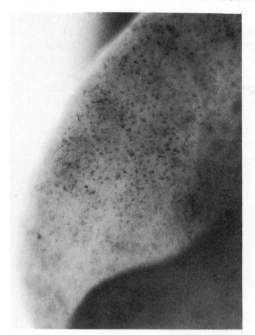

Figure 13.15 Chick embryo, 5 days of incubation, after the administration of 100 μg cyclophosphamide at 4 days. Supravital staining with Nile blue, 1:10000. Intense macrophage reaction in the dorsal integument

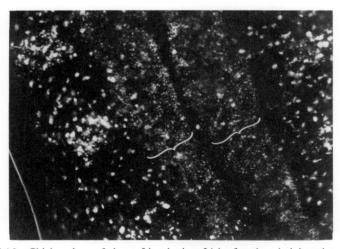

Figure 13.16 Chick embryo, 5 days of incubation. 24 h after the administration of 100 μg cyclophosphamide. Vital fluorochroming with acridine orange. Sacral region (compression). In the neural tube fine necrotic granules (}), slight macrophage reaction. In the surrounding mesenchyme intense macrophage reaction

Applying the fluorescence method to embryos treated with 100 μg cyclophosphamide 24 h before, we obtained a global picture of the cytotoxic effect within all embryonic tissues and organs. In spite of the cytotoxic effect and of the polyploidy induced, the substance did not prevent the macrophage reaction occurring, first of all in the mesenchymal areas and almost absent within the central nervous system (where only primary necrotic changes with intra- and extracellular necrospherules were present) (Figure 13.16).

A comparison between the supravital Nile blue staining and the results obtained by vital fluorochroming supports the value of our method which demonstrated the whole complex of cyclophosphamide induced cytotoxic

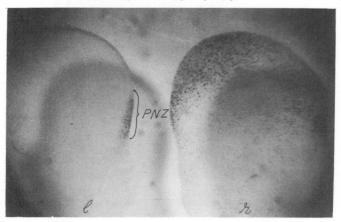

Figure 13.17 Chick embryo, 5 days of incubation. 24 h after the administration of 100 μg cyclophosphamide. Supravital staining with Nile blue. Limb. Left: control, right: treated

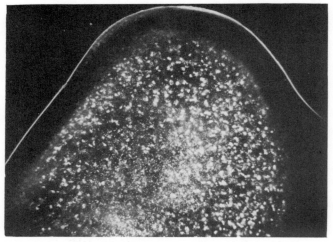

Figure 13.18 Chick embryo, 5 days of incubation. 24 h after the administration of 100 μg cyclophosphamide. Vital fluorochroming with acridine orange (slight compression). Limb. The cytotoxic effect is demonstrated by the elective fluorescence of all the intra- and extracellular necrotic elements

effects. Necroses in the central nervous system, in the liver, in the mesonephros, in the heart and other even deeply located embryonic tissues appeared as a complete, tridimensional picture. In this context it is of interest to

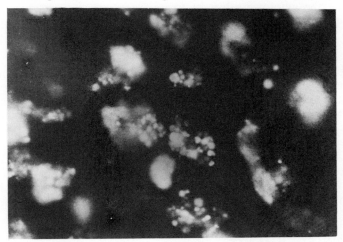

Figure 13.19 Chick embryo, 5 days of incubation. 24 h after the administration of 100 μg cyclophosphamide. Vital fluorochroming with acridine orange. Intense macrophage reaction in the mesenchyme. Marked fluorescence of phagosomes within the macrophages. Higher magnification of an area of the preparation presented in Figure 13.18

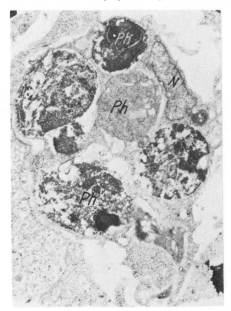

Figure 13.20 Chick embryo, 5 days of incubation. Limb mesenchyme. 24 h after the administration of 100 μg cyclophosphamide. Electronmicroscopical picture of a macrophage. N = nucleus, Ph = phagosomes. × 4800

compare the Nile blue-stained and fluorochromized limbs (Figures 13.17, 13.18, 13.19 and 13.20).

Fluorochroming may be developed as a comparative screening method for evaluating the cytotoxic effect of various antitumour drugs by using the quantitative evaluation of the macrophage reaction.

FINAL REMARKS

The method for the study of cell death in embryos by vital fluorochroming has an essential point: the administration of fluorochromes to living embryos and supravital controls. Thus, a general survey of necrosis and necrobiosis within the embryo is obtained, which is not possible by other commonly used methods. The method described facilitates a global detection of the macrophage reaction and its quantitative assessment.

One may speculate, whether the accumulation of fluorochromes (or of some other vital dyes, such as neutral red, Nile blue) can be considered an experimental model for investigating the selective effect of some teratogens in areas of normal necrosis which are subsequently transformed into teratogenic zones.

ACKNOWLEDGEMENTS

The authors are indebted to Mrs M. Cosma for technical assistance, to Mrs M. Gherman for the photographical work, to Mrs V. Fluture and physicist H. Wolff for the preparation of the electronmicrographs, and to Mrs S. Antal for the preparation of the manuscript.

References

1. Ernst, M. (1926). Über Untergang von Zellen während der normalen Entwicklung bei Wirbeltieren. *Z. Anat. Entw. Gesch.*, **79**, 228
2. Glücksmann, A. (1951). Cell death in normal vertebrate ontogeny. *Biol. Rev.*, **26**, 59.
3. Menkes, B., Litvac, B. and Ilieş, A. (1964a). Spontaneous and induced cell degeneration in relation to teratogenesis. *Rev. Roum. Embryol. Cytol.-Série Embryol.*, **1**, 47
4. Menkes, B., Deleanu,. M. and Ilieş, A. (1965). Comparative study of some areas of physiological necrosis at the embryo of man, some laboratory mammals and fowl. *Rev. Roum. Embryol. Cytol.-Série Embryol.*, **2**, 161
5. Saunders, J. W., Gasseling, M. T. and Saunders, L. (1962). Cellular death in morphogenesis of the avian wing. *Dev. Biol.*, **5**, 147
6. Saunders, J. W. and Gasseling, M. T. (1959). Effects of reorienting the wingbud apex in the chick embryo. *J. Exp. Zool.*, **142**, 553
7. Saunders, J. W. Jr. (1966). Death in embryonic systems. *Science*, **154**, 604
8. Saunders, J. W. Jr. and Fallon, J. F. (1967). Cell death in morphogenesis. In: M. Locke (ed.). *Major Problems in Developmental Biology*, pp. 289–314. (New York: Academic Press Inc.)
9. Chèvremont, M. and Chèvremont-Comhaire, S. (1945). Recherches sur le déterminisme de la transformation histiocytaire. *Acta Anat. (Basel)*, **1**, 95
10. Deleanu, M. and Prelipceanu, O. (1975). Researches on the macrophage reaction in embryonic necrotic foci by means of cell cultures. *Rev. Roum. Morphol. Embryol. Physiol. Morphol. Embryol.*, **21**, 273
11. Menkes, B. and Deleanu, M. (1964b). Leg differentiation and experimental syndactyly in chick embryo. II. Experimental syndactyly in chick embryo. *Rev. Roum. Embryol. Cytol.-Série Embryol.*, **1**, 69

12. Menkes, B., Alexandru, C. and Checiu, I. (1974). Janus grün B im teratologischen Experiment. *Rev. Roum. Morphol. Embryol. Physiol., Morphol.-Embryol.*, 19, 141
13. Pautou, M. P. (1968). Rôle déterminant du mésoderme dans la différenciation spécifique de la patte de l'oiseau. *Arch. Anat. Microsc. Morphol. Exp.*, 57, 311
14. Kieny, M. and Pautou, M. P. (1970). Sur le mécanisme de la nécrose morphogène dans le pied de l'embryon de poulet. *C.R. Acad. Sci. Paris*, 270, 3091
15. Pautou, M. P. and Kieny, M. (1971a). Sur les mécanismes histologiques et cytologiques de la nécrose morphogène interdigitale chez l'embryon de poulet. *C.R. Acad. Sci. Paris*, 272, 2025
16. Pautou, M. P. (1975). Morphogenèse de l'autopode chez l'embryon de poulet. *J. Embryol. Exp. Morphol.*, 34, 511
17. Ilieş, A. (1967). La topographie et la dynamique des zones nécrotiques normales chez l'embryon humain. *Rev. Roum. Embryol. Cytol.-Série Embryol.*, 4, 51
18. Menkes, B. (1968). Localised necrosis and necrobiosis in teratogenesis. *Rev. Roum. Embryol. Cytol.-Série Embryol.*, 5, 139
19. Zwilling, E. (1964). Controlled degeneration during development. In: A. V. S. de Reuck and J. Knight (eds.). *Ciba Foundation Symposium: Cellular injury*, pp. 352–368. (Boston: Little, Brown and Co.)
20. Menkes, B., Sandor, S. and Ilieş, A. (1970). Cell death in teratogenesis. In: D. H. M. Woollam (ed). *Advances in Teratology*, Vol. IV, pp. 169–215. (London: Logos Press)
21. Töndury, G. (1962). *Embryopatien. Über die Wirkungsweise (Infektionsweg und Pathogenese) von Viren auf den menschlichen Keimling* (Berlin, Göttingen, Heidelberg: Springer Verlag)
22. Pexieder, T. (1975). Cell death in the morphogenesis and teratogenesis of the heart. In: A. Brodal, W. Hild et al. (eds.). *Advances in Anatomy, Embryology and Cell Biology*, 51, pp. 1–100. (Berlin, Heidelberg, New York: Springer Verlag)
23. Kreybig, von Th. (1968). *Experimentelle Praenatal Toxikologie*. (Aulendorf i. Württ.: Editio Cantor K. G.)
24. Sandor, S. (1968). The influence of ethyl-alcohol on the developing chick embryo. II. *Rev. Roum. Embryol. Cytol.-Série Embryol.*, 5, 167
25. Bellairs, R. (1961). Cell death in chick embryos as studied by electron microscopy. *J. Anat.*, 95, 54
26. De Duve, C. (1975). Les Lysosomes: un nouveau groupe de granules cytoplasmatiques. *J. Physiol. (Paris)*, 49, 113
27. Forsberg, J. G. and Källén, B. (1968). Cell death during embryogenesis. *Rev. Roum. Embryol. Cytol.-Série Embryol.*, 5, 91
28. Wendler, D. (1972). Der embryo-fetale Zelltod während der Normogenese und im Experiment. In: G. Uschmann (ed.). *Acta Historica Leopoldina*, 8, pp. 7–295. (Leipzig: Johann Ambrosius Barth)
29. Litvac, B. and Litvac, E. (1973). Early ultrastructural changes during the morphogenetical necrosis of the mesenchyme in the interdigital membrane of the chick embryo. *Rev. Roum. Embryol., Cytol.-Série Embryol.*, 10, 137
30. Pautou, M. P. and Kieny, M. (1971b). Effet du vert Janus sur la nécrose morphogène interdigitale chez le poulet; analyse histologique et cytologique. *C.R. Acad. Sci. Paris*, 272, 2378
31. Milaire, J. (1965). Aspects of limb morphogenesis in mammals. In: R. L. DeHaan and H. Ursprung (eds.). *Organogenesis*, pp. 283–300. (New York and London: Holt, Rinehart & Winston, Inc.)
32. Ballard, J. and Holt, S. J. (1968). Cytological and cytochemical studies on cell death and digestion in the foetal rat feet: the role of macrophages and hydrolytic enzymes. *J. Cell Sci.*, 3, 245
33. Menkes, B., Deleanu, M., Prelipceanu, O. and Căpălnăşan, I. (1977a). Abfang, Konzentration und Lokalisation normogener oder teratogener Substanzen durch Nekrosezonen im Embryo. *Rev. Roum. Morphol. Embryol. Physiol., Morphol.-Embryol.*, 23, 27
34. Menkes, B., Prelipceanu, O. and Căpălnăşan, I. (1977b). Fluoreszenzmikroskopische Untersuchung embryo-fötaler Zelldegenereszenzen. *Rev. Roum. Morphol. Embryol. Physiol., Morphol.-Embryol.*, 23, 247
35. Ellinger, Ph. and Hirt, A. (1932). Eine Methode zur Beobachtung lebender Organe mit

stärksten Vergrösserungen im Lumineszenzlicht. *Abderhaldens Hdbch. Biol. Arbeitsmethoden*, 5, 2
36. Saunders, A. M. (1964). Histochemical identification of acid mucopolysaccharides with Acridine Orange. *J. Histochem. Cytochem.*, 12, 164
37. Hamburger, V. and Levi-Montalcini, R. (1949). Proliferation, differentiation and degeneration in the spinal ganglia of the chick embryo under normal and experimental conditions. *J. Exp. Zool.*, 111, 457
38. Levi-Montalcini, R. (1950). The origin and development of the visceral system in the spinal cord of the chick embryo. *J. Morphol.*, 86, 253
39. Wechsler, W. (1967). Electron-microscopy of the cytodifferentiation in the developing brain of chick embryos. In: R. Hassler and H. Stephan (eds.). *Evolution of the Forebrain*, pp. 213–224. (New York: Plenum Press)
40. Jacobson, M. (1970). *Developmental Neurobiology*. (New York, Chicago, San Francisco: Holt, Rinehart and Winston, Inc.)
41. Menkes, B. and Tudose, O. (1969). Determination of the emigration and tectogenesis in the embryonic central nervous system by oxygen and metabolite gradients. *Rev. Roum. Embryol. Cytol.-Série Embryol.*, 6, 159
42. Feeney, J. F. and Watterson, R. L. (1946). The development of the vascular pattern within the walls of the central nervous system of the chick embryo. *J. Morphol.*, 78, 231
43. Tudose, O. (1971). *Development of the Vascular System in the Neural Tube in Normal and Some Pathological Conditions*. Dissertation for D. Sc. (in Rumanian). (Timişoara: Library of Medical School)
44. Menkes, B., Prelipceanu, O. and Căpălnăşan, I. (1979). Perichordale Nekrosan. Ein Beitrag zum Verständnis der Morphogenese der Wirbelsäule. Rev. roum. Morphol. *Embryol. Physiol., Morphol.-Embryol.* (In press)
45. Sandor, S. and Amels, D. (1973). Multiphase analysis of the prenatal noxious action of some chemical compounds. In: E. Klika (ed.). *Acta Univ. Carolinae Medica Monographia*, LVI–LVII, pp. 117–119. (Praha: Universita Karlova)
46. Holtzer, H. and Detwiler, S. R. (1953). An experimental analysis of the spinal column. *J. Exp. Zool.*, 123, 335
47. Holtzer, H. and Mayne, R. (1973). Experimental morphogenesis. The induction of somitic chondrogenesis by embryonic spinal cord and notochord. In: R. V. D. Perrin and M. J. Finegold (eds.). *Pathology of Development or Ontogeny revisited*. pp. 52–65. (Baltimore: Williams and Wilkins)
48. Kosher, R. A. and Lash, J. W. (1975). Notochordal stimulation of in vitro somite chondrogenesis before and after enzymatic removal of perinotochordal materials. *Dev. Biol.*, 42, 362
49. Lash, J. W., Glick, M. C. and Madden, J. W. (1964). Cartilage induction in vitro and sulfate activating enzymes. *Nat. Cancer Inst. Monogr.*, 13, 39
50. Strudel, G. (1953). Influence morphogene du tube nerveux et de la chorde sur la différenciation de la colonne vertèbrale. *C.R. Soc. Biol. Paris*, 147, 132
51. Strudel, G. (1962). Induction de cartilage in vitro par l'extrait de tube nerveux et de chorde de l'embryon de poulet. *Dev. Biol.*, 4, 67
52. Zilliken, F. (1967). Notochord induced cartilage formation in chick somites. Intact tissue versus extracts. In: E. Hagen, W. Wechsler and P. Zilliken (eds.). *Experimental Biology and Medicine*, Vol. I, pp. 199–212. (Basel, New York: S. Krager)
53. Hall, B. K. (1977). Chondrogenesis of the somitic mesoderm. In: A. Brodal, W. Hild and all. (eds.) *Advances in Anatomy, Embryology and Cell Biology*, 53, pp. 1–49. (Berlin, Heidelberg, New York: Springer Verlag)
54. Tage, K. N. and Finnegan, C. V. (1970). The distribution of glycosaminoglycans in the axial region of the developing chick embryo. I. Histochemical analysis. *J. Exp. Zool.*, 175, 221
55. Menkes, B. and Checiu, I. (1979). Sklerotome, Sklerotomite und Mukopolysaccharide. (In preparation)
56. Theiler, K. (1958). Zelluntergang in den hintersten Rumpfsomiten bei der Maus. *Z. Anat. Entw. Gesch.*, 120, 274
57. Theiler, K. (1959). Schwanzmutanten bei Mäusen. Ein Beitrag zur Entstehung von Wirbelfehlern. *Z. Anat. Entw. Gesch.*, 121, 155

58. Webster, D. A. and Gross, J. (1970). Studies on possible mechanisms of programmed cell death in the chick embryos. *Dev. Biol.*, 22, 157
59. Fallon, J. F. and Saunders, J. W. (1968). In vitro analysis of the control of cell death in a zone of prospective necrosis from the chick wing bud. *Dev. Biol.*, 18, 553
60. Strudel, G. (1967). Some aspects of organogenesis of the chick spinal column. In: E. Hagen, W. Wechsler and P. Zilliken (eds.). *Experimental Biology and Medicine.* Vol. I., pp. 183–198. (Basel, New York: S. Karger)

Index

ablepharia, 99, 100
abortus and mosaicism, 4
achondroplasia, 39
acridine orange, 223
ageing of spermatozoa, 1–10
alcohol, and perinatal development, 209
amino-acetonitrile, 41
β-aminopropionitrile, 41
AMP, cyclic, levels in palate development, 169
aneuploidy, 6
artificial insemination in cattle, 9
aspirin, 38
atropine, 151, 152
L-azetidine carboxylic acid (LACA), 38

bethanechol, 151, 152
birth defects and sperm storage, 9
blastocyst
    abnormalities and sperm age, 2, 3, 8
    alkaloid treated, 51–68
    cytochalasin B and formation, 74
    preimplantation, 51–68
    pulse labelling, 53–7
    transformation from morula, 79
    ultrastructure, 62–8
brachyury syndrome, 73
Brinster's medium, 57
    modified, 52
5-bromo-2-deoxyuridine, 16–25
    chromosomal alteration, 20
    embryotoxicity, 19, 20

cadmium, 37
carbachol, 151
carbohydrate metabolism, perinatal changes, 195–8

catecholamines, neuroactive, 38
chimerism, 4, 51
chondrocytes, 36
chromatid sister exchanges, 17, 22, 23
chromosome
    abnormalities, 2–6, 8, 179, 180
    breaks, 21
    Giemsa (G)-bands, 17, 20
    heteroploidy, 21
    isolation, 16
    lateral asymmetry, 21–3
    structure and teratogens, 14–25
cleavage errors, 3, 4, 9
cleft palate, 19, 165
colchicine, 40, 51, 52, 59, 61, 68
    photo-oxidation, 63
collagen
    chemical biology, 30
    enzymes and synthesis, 31–3
    phenotype changes, 34
    role in tissue differentiation, 29, 30
    secretion inhibition, 39
    types, 30, 31
collagen synthesis, 31
    altered rates, 34–6
    and teratogens, 33–42
concanavalin A receptors, 73
copper deficiency, 41
corticosterones, 200
    and differentiation, 207
coxsackie B-4 virus, 185, 186
curare, 151, 152
cyclic AMP, 169
cyclophosphamide, 95–114, 137
    and ectoderm derived tissue, 103–7
    -induced fetal abnormalities, 95–114
    and mesoderm derived tissue, 106–8

cytochalasin B, 31, 40, 71–90, 150
  and embryogenesis, 71–3
  mode of action, 86
  and preimplantation embryos, 74, 75, 77–9, 86–8
  and postblastocyst embryos, 75–7, 79–90
cytogenetic abnormalities, and sperm age, 1-11
cytosine arabinoside, 137

diabetes, experimental, 39
deoxyribonucleoprotein (DNP), 10
6-diazo-5-oxo-L-norleucine, 33–5, 167
digyny, 3
diphenylhydantoin, 38
  and perinatal development, 209
dispermy, 3
disulphide content of spermatozoa, 10
Down's syndrome, 183
drugs, cytotoxic, 137
dystopia, 101

ectrodactyly, 19
embryo
  [^{14}C]glucosamine uptake, 54–68
  mortality, 3
  toxicity, 19, 20
embryogenesis, 71–4
encephalocele, 99, 102
enzymes
  carbohydrate, 195–200
  and collagen synthesis, 31–3
  extra-uterine changes, 202
  inhibition and collagen synthesis, 33
  lipid, 199, 200
  perinatal changes, 195–8
epidermal growth factor, 163
exophthalmia, 99, 100
extrauterine life
  factors modifying, 206–11
  impaired adaptation, 193–211

fertility and sperm storage, 5
fertilization errors, 3
fetal abnormalities, cyclophosphamide-induced, 95–114
fetogenesis, 96, 97
fetus, development in rat, 97, 98
fibrillogenesis, 31, 39
5-fluorodeoxyuridine, 137
foyer primaire preaxial ('fpp'), 137–40

galactosyl transferase, 31, 32
gastrochisis, 101
Giemsa (G) bands, 17, 20
glucagon, 200, 201
  and differentiation, 208, 209

glucocorticoids and palate development, 162–166
[^{14}C]glucosamine, 51–68
glucosyl transferase, 31, 32
growth retardation, intrauterine, 203

heart development, and retinoic acid, 119–33
Helminthosporium dermatioideum, 71
herpes simplex virus, 188
hexamethonium, 151
hormones
  and growth retardation, 204
  perinatal changes, 200–2
5HT, 151–5
hyaloplasmic fan, 85
hydralazine, 38
hydronephrosis, 101
hypoglycaemia at birth, 204

ICM, 73, 75, 81–5, 87, 88
  ectopic, 79
implantation, effects of cytochalasin B, 71–90
inhibitors, metabolic, and collagen synthesis, 33
insulin, 200, 201
isofamide, 95

karyolysis, 103
karyopycnosis, 103
karyorhexis, 103

LACA (L-azetidine carboxylic acid), 38
lactate dehydrogenase, 36
lathyrogens, 41
limb development
  cell death in, 135–141
  see also necrosis
lipid metabolism, perinatal changes, 199, 200
lumicolchicine, 59, 61, 63, 68
luminal, and perinatal development, 209
lysyl hydroxylase, 31, 32
lysyl oxidase, 31, 32, 41

malate dehydrogenase, 36
malformations *see* fetal abnormalities
manganese, 39
6-mercaptopurine, 137
mercury
  methyl, 37
  teratogenicity, 37
methysergide, 151
micrognathia, 99
microtubule disruption, 40, 61, 63
microvilli, 65–7
mitomycin C, 16–25
  -induced chromosomal alterations, 21, 22
Moloney murine leukaemia virus, 189, 190

INDEX

morula-to-blastocyst transformation, 79
mosaicism, 4, 6
  aneuploid, 4
  euploid, 4
  N-mustard, 95
myocytes, effect of retinoic acid, 125–7
myosin ATPase, 144, 145

necrosis
  cell staining, 136
  interdigital zone, 140, 141
  physiological, in rodent limb, 135–41
necrotic cells
  detection methods, 221–5
  normal CNS, 225–8
  vital fluorochrome and, 219–38
neurotransmitters, and palate morphogenesis, 151–7
nitrosourea
  N-ethyl, 95
  N-methyl, 95
non-disjunction, 9
nuclear fragmentation, 9

oligodactyly, 101
ovulation induction, 4

palate
  analysis, 144
  cleft, 19, 165
  contractile systems, 148
  mesenchyme culture, 153, 154
  morphogenesis model, 156, 157
  primary development, 169–72
  rotation, 148–51
    and neurotransmission, 151–5
  secondary development, 161–70
  shelf morphology, 148, 149, 156
palate morphogenesis, 143–57
  and macromolecule synthesis, 161–73
palladium, 38
parvovirus, 183
phagocytosis, myocardial cell, 128
polydactyly, 137
polymorphism, C-band, 3
polyoma virus, 182
polyploidy, 3, 72, 90
postmaturity, and metabolism, 205, 206
prematurity, and metabolism, 205
procollagen
  cleavage, 40, 41
  peptidase, 31
prolyl hydroxylase, 31, 32
  inhibition, 37, 38
  and morphogenesis, 37
protein biochemistry, 143
protein, contractile, and palate morphogenesis, 143–57

protein metabolism, perinatal changes, 198
pyridostigmine, 151, 152

radioautography, 62, 65
Rammskopf, 100
Rana pipiens, 71, 86
retinoic acid
  and cell budding, 125
  and heart development, 119–33
Rous sarcoma virus, 35
rubella syndrome, 178–82

salicylates, and perinatal development, 209
scanning electron microscopy, 52, 66
semicarbazide, 41
  thio-, 41
serotonin (5HT), 151–5
  and cell contractility, 154
simian-40 virus, 188, 189
spermatozoa
  ageing, 1–11
  defects, 2
  deoxyribonucleoprotein (DNP), 10
  disulphide content, 10
  in utero storage, 3–5, 10
  in vitro storage, 5–9
  storage temperature, 5–9
spina bifida, 101
steroids, and collagen synthesis, 33
superovulation, 52
syndactyly, 19

temperature, and sperm storage, 5–9
teratogens, 71–90, 95–114, 119–33
  biochemical mechanisms, 33
  and chromosome structure, 14–25
  and collagen synthesis, 29–42
  intrauterine growth retardation, 203–5
  see also cyclophosphamide, cytochalasin B, retinoic acid
tetraploidy, 72
thiosemicarbazide, 41
thymidine substitution, 18
thyroid function, perinatal changes, 201, 202
thyroid hormones, 208
triploidy, 3, 6, 8, 9
trisomy, 2, 3
trophoblast outgrowth, 75–90
tunicamycin, 31
turbidities, 100

ultrastructure, blastocyst, 62–8

vinblastine, 40, 52, 59–68
  and glucosamine uptake, 59–64
virus
  and collagen synthesis, 33
  coxsackie B-4, 185, 186

virus (*continued*)
   cytocidal, 187, 188
   H-1, 183
   herpes simplex, 188
   Moloney murine leukaemia, 189, 190
   oncogenic, 188–90
   parvovirus, 183
   polyoma, 182
   and preimplantation embryos, 185–8
   and prenatal disorders, 177–90
   Rous sarcoma, 35
   rubella
      cataract formation, 181
   effect on bone growth, 179, 180
   effects on chromosomes, 179
   and mitosis, 178, 179
   simian-40, 188, 189
   susceptibility and differentiation, 182–90
vital fluorochroming
   and embryo cell necrosis, 219–38
   and study of prevertebral axial organs, 228–38

zinc, 37, 38
zona pellucida, and drug permeability, 68
zonula adherens, 145